The Man Who Tasted Shapes

The Man Who Tasted Shapes

▲ ■ ●

RICHARD E. CYTOWIC, M.D.

with a foreword by Jonathan Cole, M.D.

A Bradford Book
The MIT Press
Cambridge, Massachusetts
London, England

MIT Press edition with new afterword, 2003
First MIT Press edition, 1998
First published in 1993 by G. P. Putnam's Sons, New York
A Jeremy P. Tarcher/Putnam Book.

Printed and bound in the United States of America

Library of Congress Cataloging-in-Publication Data

Cytowic, Richard E.
   The man who tasted shapes / Richard E. Cytowic.
      p.   cm.
   "A Bradford book."
   Originally published: New York: Putnam, © 1993.
   Includes bibliographical references and index.
   ISBN 0-262-53255-7 (pbk.: alk.paper)
   1. Synesthesia. 2. Emotions and cognition. I. Title. [DNLM: 1. Taste–physiology.
2. Perception. 3. Cognition. 4. Emotions. Not Acquired / WI 210 C997m 1993a]
RC394.S93C96   1998
152–dc21
DNLM/ DLC   97-38214
for Library of Congress   CIP

UK and Eire edition published by Imprint Academic, ISBN 0907845 436

10 9 8 7

To Bobby, in memoriam,
and in memory of
Reverend Clark A. Thompson and Michael O. Watson

Contents

▲ ■ ●

Part Two
ESSAYS ON THE PRIMACY OF EMOTION

Foreword to the MIT Press Edition

▲ ■ ●

JONATHAN COLE, M.D.

"Memory Holes"

Science, like the world of fashion, has fashions. Some ideas and theories enjoy enormous interest for a while and then fade from view. Physiognomy, the "science" of extracting character from facial features was enormously, and erroneously, influential in the nineteenth century. Lavater's book on the subject sold hugely for its day,[1] and so influential was it that Darwin was nearly refused a berth on *The Beagle* because his nose was too long. One might not be surprised that such ideas are popular for only a short time; more curiously, some of the very phenomenology upon which science depends has also been shown to appear and then disappear like Alice's cat. Oliver Sacks[2] discusses several of these in a recent essay. Tourette's Syndrome, the "Geometrical Spectra" of migraine, and the alienation of limbs after injury were all described and then lost in the so-called "memory holes," a term coined by Orwell.[3] The latter condition, although described by Weir Mitchell[4] within the context of American Civil War, has still not been investigated thoroughly.

Synesthesia is another neurological abnormality that has been in and out of fashion. First described by Locke in 1690[5] and medically by Woolhouse twenty years later, it was the subject of much scientific interest a hundred years ago. By the 1940s, however, it had virtually disappeared from the collective scientific and medical consciousness. Cytowic[6] and Baron-Cohen et al.[7] agree that with the rise of behaviorism as the preëminent neuroscientific discipline, synesthesia, which depended on individual introspection, was no longer considered worthy, or even possible, to study. In synesthesia an individual experiencing a sensation in one sensory modality also experiences, involuntarily, a sensation in another sensory modality. The most common experience seems to be seeing color when hearing sounds. This cross-modal sensation is reproducible in a given individual during their lifetime, so that a given sound or word always leads to perception of the same color. Because of its curious phenomenology and its failure, usually, to lead to significant problems in daily living, its existence has appeared hidden from medicine, including neuroscience, for the latter part of this century.

That now the subject is both known and researched is due in no small part to Richard Cytowic's meeting with a neighbor, Michael Watson, who, by chance, let slip that the chicken he was cooking tasted "round" rather than "pointed." Many neurologists would have treated this comment with a detached and temporary curiosity before moving on, especially if not "on duty." For Cytowic, by contrast, it was the start of an extraordinary journey to understand that "shaped taste" of synesthesia and to speculate about its neurological basis, a journey he relates with candor and considerable personal intellectual transparency in *The Man Who Tasted Shapes*. Within this book is an important exploration of the detailed phenomenology of the condition based on Cytowic's careful observation and experimentation with synesthetics over a period of a decade or so. *The Man Who Tasted Shapes* is not, however, the product of a research or academic program but the fruits of one man's fascination and, indeed, necessary obsession that develop as he pursues his career in neurology.

The book gives details of two cases of synesthesia—Watson's and Victoria's—and discusses some of the consequences. In one fascinating passage, for instance, Cytowic and Watson discuss cooking and how his synesthesia enhances his creativity and enjoyment: "He never followed a recipe but liked to create a dish with an 'interesting shape.' Sugar made things taste 'rounder,' while citrus added 'points.' Michael continued, 'The shape changes with each moment, just as flavor does. . . . French cooking is my favorite precisely because it makes the shapes change in fabulous ways.'"

One can see why it was important to discriminate the synesthetic sense from the metaphorical use of language that so
appealed to the many artists of the early twentieth century, who,
Cytowic details, were fascinated by the condition. This he covers
in the book's more scientific parts but also, immediately, he raises
the matter with Michael, who replied, "'. . . for me, wine really has
a shape. Describing some wine as "earthy" is not poetic because it
can be literally like holding a clod of dirt in my hand.'"

We actually learn little of Michael's personal life, despite the evident friendship between him and Cytowic. I was very fortunate
that all but two reviews of my first book,[8] a neurological biography, were favorable. One critic complained that she had not
learned enough about Ian, the subject, and of how his character
and personality had changed because of his problem. I had, in
fact, considered such a portrait when writing the book but backed
away. The art of neurological biography requires movement of a
relationship from doctor/patient to author/subject. But more than
that, if the two become friends, then certain aspects of a life are
off limits. Cytowic's reluctance to tell us more than is necessary
neurologically reflects his integrity.

Rollercoaster Ride

So, although this is not a full biography of the man who tasted
shapes, it does become a double biography. For as we learn much,
neurologically, about Michael and about his synesthesia, we are
also drawn, uniquely, into an intellectual autobiography of
Richard. We learn much of Cytowic's thought processes, hunches,
bloody mindedness, and persistence in uncovering the story and
the condition over a period of a decade or so. Along the way we
also learn of the indifference, and at times hostility, of some doctors to his pursuing the subject at all; we learn something about
the nature of medicine and medical education in America at this
time and of Cytowic's opinion of it. In revealing the nature of his
scientific investigations, the book is outstanding. It then goes further to lay open his struggles to understand the condition and
place it within a wider and novel theory of brain action. In this it
is a most ambitious work, for whatever else he does Cytowic never
hides his feelings, or his prejudices, making for a rollercoaster of a
book. At times, however, this rollercoaster gathers rather too
much momentum, as rollercoasters are prone to do.

Cytowic's develops a theory of synesthesia being dependent on a part of the brain, the limbic system, which lies beneath the cerebral cortex, and in doing so rejects a cortical site for some of the abnormalities in synesthesia. This he did on the basis of an experiment in the early 1980s: It may prove premature. The technique he was forced to use measured cortical blood flow but could not measure limbic sub-cortical blood flow, and has been superseded by more sophisticated and accurate techniques that can measure global intracerebral metabolism. With newer methods it has been shown that when synesthetics "see sounds" they do, in fact, use areas of the cortex dedicated to higher visual function.[9] This work, of course, could not have been known when Cytowic began to develop his thoughts about brain function, but may, in time, lead to acceptance of a different model. Another finding, however, from the original experiments, of reduced cortical metabolism, has received some support from these newer techniques, though its significance remains unclear.

Cytowic went on from his limbic hypothesis to develop, with force and enthusiasm, ideas about the primacy of emotion in individual functioning in a manner that downgraded cortical function and cognition. In this, I suspect, he took a deliberately argumentative and idiosyncratic view, stimulating others to consider their position. For most people in the field, the anatomical structures of the brain (the insula, cortex, hippocampus, and so on) work together as a whole, just as emotional and cognitive factors coexist and interact. There is now also an increasing realization that the isolated study of small aspects of brain function, though enormously informative, has its limitations. With PET, MRI and fMRI, brain action as a whole can be observed almost non-invasively. Cytowic's development of ideas is one of the most fascinating aspects of this book. He is, I think, aware of some of the excess momentum within the rollercoaster ride, and he discusses some of these issues and points readers toward more recent work in the new afterword to the book.

"Not Nuts"

One might think of the journey to understand synesthesia as being of academic or even literary interest, and of course it is. Arguably, more important is the recovery of the phenomenology to our medical consciousness and in particular to a general awareness. Cytowic's writings and appearances in the media have educated millions of people about synesthesia. But more than that, as he explains in the book, many previously isolated synesthetics have been made aware of their condition, that it has a name and that "they are not nuts." Most people with neurological problems wonder if it is all in the mind—imagined and unreal. To be shown this is not the case is hugely reassuring and therapeutic in a way that cannot be underestimated. For his part in this Cytowic deserves enormous credit. It is given to few of us in medicine to help people whom we have never met.

This educational process continues, and Cytowic, in his afterword, gives several web sites where synesthetics can exchange information—how different from his first encounter with the phenomenon, when he was told by a colleague, "Forget it, this is not on your neuro hit parade because it's not real."

The Nature of Synesthesia

While some of Cytowic's ideas remain controversial, his interest in the subject has helped stimulate work. The book discusses many interesting facets of synesthesia and speculates about its origin. He suggests, for instance, that it may be a normal process but one hidden below consciousness in all but a few of us. In other words, we might all be synesthetic at a subconscious level of central nervous system functioning.[10] As Cytowic suggests, synesthetics lead one to consider why sensory experience is usually modular, why touch and vision, for instance, or hearing and smell, are kept separate.[11,12] Might we all have been synesthetic at one time during development, and, if so, what might this suggest for central nervous system development?

That we might not have a nonconscious synesthesia was suggested by experiments published after Cytowic's book.[9] These found that cortical areas of convergent input between vision and sound activated in synesthetics during "seen" sounds were not active in controls. This suggests not only that the phenomenon involves cortical structures according to a predictable model but also that controls do not activate these areas of convergence during sound stimuli without synesthetic visual perception. In addition, as Cytowic points out, synesthetics do not have generalized nonmodularity of sensory experience, but rather, have specific patterns of perception, with particular and unaltering sounds leading, say, to similar visual perceptions that are reproducible and also idiosyncratic in a manner unlikely ever to be useful (French cooking apart).

It is clear from several examples within *The Man Who Tasted Shapes*, that for Michael and others their experiences of synesthesia were on the whole pleasurable, though not apparently so pleasurable that they became abused. Michael's enjoyment of his synesthesia seems no greater than Richard's reminiscences of his own childhood with his mother's oil-paints. If not pleasurable then might synesthesia lead to confusion about sensations within individuals? This does not usually appear to be the case, though in one case, a musician with unusual synesthesia both ways (from color to sound and vice versa), it was reported that she did experience stress and uncertainty from this tangling.[7] Her experiences suggest the importance of distinct sensations, so that we can distinguish sounds from sights.

There do remain several indirect pieces of evidence that gnaw at the edges of modularity of sensation. There are many well known illusions that depend on cross-modal sensory experience: for instance, a large object appears to weigh more than a small one of a similar weight; our expectations and perceptions of one sensation, heaviness or effort, can be unconsciously influenced by expectations via another sensation, vision. These probably depend on learned experience, since in this case, large objects are usually heavier. We often form impressions in one sense that are based on a combination of senses. Thus the learning of the location of sounds in external space will be aided by visual knowledge of where the object making the sound actually is. It has recently been shown that, in some animals, development of the area of the brain responding to location of sound in external space is reduced if the animal is deprived of vision.[13] Thus some cross-modal information is used in the development of our sensory apparatus during early critical periods for learning, at least. It seems unlikely, however, that this gives a clue to the origins of synesthesia, since synesthetic perceptions, as Cytowic relates, are elemental in terms of color and vague shape, rather than specific in terms of formed visual objects or specific distances.

Further clues come from genetic studies mentioned by Cytowic in his parallel scientific book, *Synesthesia: A Union of the Senses*.[6]

There is becoming clearer evidence of the genetic basis of the phenomenon, with the mode of inheritance either being passed down
the maternal or paternal line but expressed more in females (who
outnumber males 6 to 1 for synesthesia), or, more likely, being
linked to the sex of the child in a rather complex manner.[7] This
points to a single gene abnormality (curiously like another "lost"
condition, Tourette's Syndrome). If this is so, then the ideas that
we all have synesthesia but are unaware of it, or that cross-modal
sensory learning is related to the phenomenon, seem most unlikely.

A central, implicit theme through the book is what, if any, relation synesthesia has with normal brain function and what therefore we can learn from it. Whether it allows Cytowic's theoretical
modeling or whether we are, in the end, left with an isolated but
fascinating perceptual difference arising from a gene mutation,
will be addressed by future work. The basis of the condition
remains tantalizingly out of reach. At the least, it allows us to
understand something of modularity and separateness of normal
sensory function and to view ourselves in a different and richer
light.

Theories and Feelings

In the second, smaller, part of the book, Cytowic leaves synesthesia behind to develop a theory of the primacy of emotion over
those strictly cognitive aspects of the self. He clearly states that
this part represents a trial or attempt to explore themes of rationality, emotionality, and conscious agency.

The role of the limbic system, the importance of emotional
valence in our responses to the world and in motivation, artificial
intelligence, the nature of language and of metaphor, and consciousness itself are all considered in a wide-ranging and idiosyncratic series of short essays. At times reading this part is like watching a giant helium balloon, as Cytowic, full of ideas, boldly develops arguments and theories, straining for lift off as he aims for the
higher view. Sometimes one wishes for more factual ballast, but
there is no denying it is a heady ride. Though the journey is uncon-
ventional, and frequently vertiginous, the destination reveals an
elemental truth. For, finally, the ride comes to rest with a plea to
focus more on our emotional being. In this he anticipates several
recent attempts by others, including Damasio[14] and myself,[15] to

place emotion within a theoretical framework of personal action and identity.

It is clear that synesthesia has fired Cytowic's scientific and intellectual curiosity, but this, for him, was not enough. Early in the book he describes how he was brought up between the two worlds of his father's medicine and his mother's art. He describes how he has remained torn between a scientific and an artistic career, between the slow and frequently tedious deliberations of theory proving and the bold expansive and imaginative brush strokes as paint hits canvas.

To change the metaphor, Cytowic's book bravely and, at times, brilliantly reveals itself as being like two horses straining in different directions: the careful, quiet, educated dressage horse and the wilder, fiery hunter. Few neurological narratives are this candidly autobiographical, and few promise such an exhilarating ride. Dear reader, you will switch from one horse to another: Enjoy the controlled pirouette as well as the gallop. But beware, these two may not always suit their rider equally: Sometimes even the most well trained horse takes off in odd directions, so involved is he with the thrill of the chase. My hope has been to enthuse you with his chase while not neglecting other, newer scientists back in the ménage, toiling away more slowly and quietly.

I began with the outlandish comparison of science and fashion. In fact, where fashion is—or should be—fast, irreverent, and careless, science is usually careful, slow, and painstaking. Once something is verified it not only endures and informs about itself but also throws light on other matters in related fields. Synesthesia is a good example of this. For his extraordinary work in rediscovering and describing synesthesia, Richard Cytowic's work has been enormously valuable. For his part in allowing those with the condition to name and understand the nature of their experience, he may justifiably feel content. This book also reveals something of where Cytowic's intellectual journey with synesthesia led him and in doing this it also reveals much about the author. As a biography of a synesthetic, description of the phenomenon, and as an autobiography of a neuroscientist, struggling to find concepts large enough to hold his ideas, the work has much to enjoy.

Hampshire, UK
August 1997

Notes

1. Lavater, Johann Caspar. 1772. *Essays on Physiognomy, Physiognomical Fragments of the Promotion of the Knowledge and Love of Mankind.*

2. Sacks, Oliver. 1995. "Scotoma: forgetting and neglect in science." In R.B. Silver (ed.), *Hidden Histories of Science*, New York: New York Review of Books.

3. Orwell, G. See Oliver Sacks, op. cit., p. 151.

4. Mitchell, Weir S. 1873. *Injuries of Nerves.* New York: Dover (reprinted in 1965).

5. Locke, J. 1690. *Essay Concerning Human Understanding,* Book 3. London: Bassett (reprinted in 1984 by Clarendon Press).

6. Cytowic, R. E. 1989. *Synesthesia: A Union of the Senses.* New York: Springer-Verlag.

7. Baron-Cohen, S., Burt, L., Smith-Laittan, F., Harrison, J., and Bolton, P. 1996. "Synesthesia: prevalence and familiality," *Perception* 25: 1073–1079.

8. Cole, J. 1995. *Pride and a Daily Marathon,* Cambridge, MA: MIT Press.

9. Paulesu, E., Harrison, J., Baron-Cohen, S., Watson, J. D. G., Goldstein, L. H., Heather, J., Frackowiak, R. S. J., and Frith, C. D. 1995. "The physiology of colored hearing: A PET activation study of color-word synesthesia," *Brain* 118: 661–676.

10. See Brown, J. W. 1988. *The Life of the Mind.* Hillsdale, NJ: Lawrence Erlbaum Associates, Inc., for a consideration of nonconscious brain functioning.

11. Fodor, J. 1983. *The Modularity of the Mind.* Cambridge, MA: MIT Press.

12. Baron-Cohen, S., Harrison, J., Goldstein, L. H., and Wyke, M. 1993. "Colored speech perception: Is synesthesia what happens when modularity breaks down? *Perception* 22: 419–426.

13. King, A. J. and Moore, D. R. 1991. "Plasticity of auditory maps in the brain," *Trends in Neuroscience* 14: 31–37.

14. Damasio, A. 1994. *Descartes' Error: Emotion, Reason, and the Human Brain.* New York: Putnam.

15. Cole, J. 1998. *About Face.* Cambridge, MA: MIT Press.

List of Illustrations and Tables

▲ ■ ●

Illustrations

Tables

Acknowledgments

▲ ■ ●

I deeply appreciate the generosity of American poet Edwin Honig in translating the Rimbaud poem *Voyelles*. Thanks also to my colleague Ayub Ommaya, M.D., whose seminar at the Smithsonian Institution on "Computers and Consciousness" was a rich source of inspiration. The late abbess of Shasta Abbey, Rōshi Jiyu-Kennett, proved that looking through different windows into the same room is not a metaphor.

Part One

▲ ■ ●

A MEDICAL MYSTERY TALE

Chapter 1

FEBRUARY 10, 1980:
NOT ENOUGH POINTS ON THE CHICKEN

"Keep me company while I finish the sauce," Michael beckoned, pulling me away from the other guests. I followed, scrutinizing the curious layout of his home. Both it and my new neighbor were pretty hip for suburban North Carolina.

His house had no inside walls. Its "rooms" poured into one another instead of keeping to well-defined spaces as rooms in most homes do. When I sat down among the appliances—what he called the kitchen—it struck me how jarring the open funkiness of a Bohemian loft was in the Bible belt. Yet I suppose it made sense, because Michael taught at the School of the Arts. Artists were supposed to be eccentric.

I quickly identified with the offbeat atmosphere of Michael's house, an attraction that stirred up an old conflict. I was supposed to wear the conservative mask expected of physicians, yet the house spoke to the eccentric and artist in me, too, a part that had to express itself with care. I was glad Michael had invited me to dinner. I had long preferred the company of creative people over that of stuffy medical types, which is why I liked living next to the conservatory.

I sat nearby while he whisked the sauce he had made for the roast chickens. "Oh, dear," he said, slurping a spoonful, "there aren't enough points on the chicken."

"Aren't enough what?" I asked.

He froze and turned red, betraying a realization that his first impression had been as awkward as that of a debutante falling down the stairs. "Oh, you're going to think I'm crazy," he stammered, slapping the spoon down. "I hope no one else heard," he said, quickly glancing at the guests in the far corner.

"Why not?" I asked.

"Sometimes I blurt these things out," he whispered, leaning toward me. "You're a neurologist, maybe it will make sense to you. I know it sounds crazy, but I have this thing, see, where I taste by shape." He looked away. "How can I explain?" he asked himself.

"Flavors have shape," he started, frowning into the depths of the roasting pan. "I wanted the taste of this chicken to be a pointed shape, but it came out all round." He looked up at me, still blushing. "Well, I mean it's nearly spherical," he emphasized, trying to keep the volume down. "I can't serve this if it doesn't have points."

An old-fashioned and odd diagnosis came to mind, but I wanted to hear more in Michael's own words to be sure. "It sounds like nobody understands what you're talking about," I finally said.

"That's the problem," sighed Michael. "Nobody's ever heard of this. They think I'm on drugs or that I'm making it up. That's why I never intentionally tell people about my shapes. Only when it slips out. It's so perfectly logical that I thought everybody felt shapes when they ate. If there's no shape, there's no flavor."

I tried not to register any surprise. "Where do you feel these shapes?" I asked.

"All over," he said, straightening up, "but mostly I feel things rubbed against my face or sitting in my hands."

I kept my poker face and said nothing.

"When I taste something with an intense flavor," Michael continued, "the feeling sweeps down my arm into my fingertips. I feel it— its weight, its texture, whether it's warm or cold, everything. I feel it like I'm actually grasping something." He held his palms up. "Of course, there's nothing really there," he said, staring at his hands. "But it's not an illusion because I feel it."

One more question, to be certain. "How long have you tasted shapes?"

"All my life," he said. "But nobody ever understands." He shrugged and carved up the chickens. "Am I a hopeless case, Doc?"

"Not at all," I answered. Just as there were no walls between the rooms of his house, I knew that Michael had no walls between his senses. Just as his rooms flowed into each other, so too taste, touch, movement, and color meshed together seamlessly in his brain. For

Michael, sensation was simultaneous, like a jambalaya, instead of neat, separate courses. Still, my self-satisfaction at recognizing one of the rarest of medical curiosities must have been perfectly clear.

Michael's glower tore me out of my reverie. "What are you grinning about?" he scolded. "I thought you would be sympathetic!"

"I'm not making fun of you," I laughed. "I'm just delighted to know someone with synesthesia. I've never met anyone who had it."

"Synthes . . ." he sputtered.

"*Syn-es-thē-sia*," I repeated. "It's Greek. *Syn* means 'together' and *aisthēsis* means 'sensation.' Synesthesia means 'feeling together,' just as *syn-chrony* means at the same time, or *syn-thesis* means different ideas joined into one, or *syn-opsis* means to see all together. You've never heard the word?" I asked.

A glow of recognition washed over Michael's face. "You mean there's a *name* for this? Is that why you're grinning?"

"Sure, and I know a little about it. People with synesthesia have their senses hooked together," I started to explain. "They can hear colors or feel sounds. Yours is—well, it looks like you taste shapes."

"What a relief!" Michael interrupted. "You mean I'm normal?"

"Normal is such a relative term. Let's just say that you're a rare bird," I suggested, "different, but not unheard of."

And with that roast chicken dinner started a research effort and a friendship that has lasted more than a decade.

Chapter 2

▲ ■ ●

THE WORLD TURNED INSIDE OUT

Only ten people in a million are like Michael Watson, born to a world where one sensation *involuntarily* conjures up another. Sometimes, all five clash together, along with a feeling of movement. That makes for six separate sensations that can mesh. What kind of world do such people live in, these *synesthetes?* How does such a world come to exist?

Imagine that you are a synesthete, like Michael Watson. You are standing in front of the refrigerator late at night trying to decide on a snack. You look at the leftover roast but say to yourself, "No, I'm not in the mood for arches." Or, contemplating a slice of lemon meringue pie, decide you aren't hungry for points. You dismiss the thought of a peanut butter sandwich because you know you couldn't sleep well if you stuffed yourself full of spheres and circles.

There you stand, bathed in the refrigerator light, casting your eye from shelf to shelf. You shift your feet against the cool floor and finally take a slice of chocolate mint pie. As you do, you feel a dozen columns before you, invisible to the eye but real to the touch. You set the fork down and run your hand up and down their cool, smooth surfaces. As you roll the minty taste in your mouth, your out-stretched hand rubs the back curve of one of the columns. What a sumptuous sensation. The surface feels cool, refreshing, even sexual in a way.

I would not pose this scenario were it not for the fact that Michael Watson has stood many times, undecided, just like this in front of his refrigerator. Other synesthetes have different experiences, each highly personal, but synesthesia itself has been known to medicine for two hundred years. The issue that synesthesia is a product of the brain (like all sensation), rather than a product of the imagination, was settled at the turn of the century. It remained both obscure and a medical mystery, however, because no one had ever been able to *explain how* the senses became garbled in this strange way.

Until now.

You will learn what goes on in the brains of these gifted people. This book, however, is about far more than the small group of people who inhabit the peculiar world of synesthesia. The solution to the medical mystery of synesthesia has profound implications for all of us. Consider, if you will, synesthesia as a diving board from which we will plunge into the engmas of the mind and discover opportunities for further developing our human potential.

Michael Watson and I first approached the puzzle of synesthesia as analysts expecting an objective answer, possibly a tangle of neurons, a short circuit that we could point to and say, "Ah ha, here's the culprit." We could not possibly have realized at the time how deep we were in an adventure that increasingly laid bare the neurological evidence for seeing the primacy of emotion over reason; the impossibility of a purely "objective" point of view; the force of intuitive knowledge; and why affirming personal experience yields a more satisfying understanding than analyzing what something "means."

Because our adventure has such rich implications, I would be dissatisfied were I to point to a dollop of brain tissue, announce "here's the spot," and then walk away. Instead, I shall explore two questions: What is the nature of synesthesia? and What is its value? I intend to explain not its meaning for the ten people in a million who have it, but the meaning of synesthesia for the 999,990 of us whom it does not affect directly.

Solving the mystery of synesthesia eventually led me to a new conception of the organization of mind that emphasizes the primacy of emotion over reason. This new concept of mind has profound implications for many fields of human endeavor—implications that I explore in the essays in *Part 2*. All because there were not enough points on a chicken many years ago, both Michael and I glimpsed into a deeper reality that exists in all individuals, but one that rarely comes to the surface of awareness. Now, you can share that inner journey.

The beginning of our journey is a medical detective adventure, one that affords a fly-on-the-wall look at how science is really done. You will see that science, far from being dominated by computers and other technologies, is really a humane enterprise with intellectual, practical, emotional, æsthetic, and moral dimensions.

As we move through the medical puzzle of synesthesia, our detective's magnifying lens will make your currently held ideas about human reason and emotion seem strange and distorted, as if what was once clear has suddenly gone out of focus. But your eyes will soon adjust and you will discover a new clarity. For example, instead of the usual recounting wherein sensation flows from the world outside *inwards* to the brain, our new view reverses the direction so that *sensation emanates from the inside out.* Your brain is an active explorer, not a passive receiver.

At the end of our journey you will have a new view of the mind and what it means to be human—a view that challenges the foundations of traditional thought and enterprise, a radical view that turns inside out and upside down conventional ideas about reason, emotion, and who we are.

Chapter 3

1957—DOWN IN THE BASEMENT:
THE MAKING OF A NEUROLOGIST

Funny how I remember those smells. Not the actual smells really, but things connected with them. The postman brought such peculiar items to a doctor's family that even the mail had a signature odor. It was typical in the 1950s for a physician's office to be in his home. In addition to mundane magazines and letters, the mailman left piles of boxes, tubes, and queerly shaped packages held together with leather straps. This was a daily occurrence, the arrival of pharmaceutical advertisements by the pound.

Enclosed with the miniature bottles of red and blue capsules, and the samples of yellow tablets sealed in cellophane, were cheap gifts with product names boldly imprinted on them. Whether these bribes imprinted themselves on my father's mind and led him to prescribe those drugs more often, I never knew, but that medicinal smell, impregnated in the knickknacks and the pills, is an indelible trace from my childhood.

Past the small library that connected the house to Father's inner sanctum lay another world bounded by another unmistakable scent, that of a doctor's office. To this day I remain perplexed as to its ingredients. I could never get my own office to smell the same, a great disappointment. Come to think of it, none of my own colleagues' offices smelled the way that those of my father's generation did. In my mind, that tincture of alcohol and disinfectant—and whatever else—conjures up contrary images. It makes me think of authority and the dread of sickness, yet a safe haven all the same.

Mostly, I associate it with the art of medicine before machines pried themselves between doctor and patient. I made my own office peaceful and inviting in a deliberate attempt to lessen the threat of imposing equipment like electrodes and our CT scanner. I struggled like an alchemist to re-create that bygone scent, but never succeeded.

From the second floor of our house came another smell, of linseed oil and turpentine. This was Mother's art studio. It was magic to see her turn a white canvas into a portrait. I was her patient assistant, waiting for her signal to squeeze out a colored worm of Grumbacher paint onto the white palette. How I hoped she would want one of the exotic names—cadmium red, raw umber, viridian green, or vermilion. Her oil bottle had been refilled so often that years of drips had turned its green label translucent. From her petite bottle she poured droplets of linseed oil onto the paint, swished a brush in the turpentine, scooped up a dab of another color, and swirled the whole mess together. Bright worms of paint squiggled across the white palette, and then plunged into pools of beautiful color.

We used a lot of white, Mother and I, in mixing colors. It took both hands to squeeze that fat two-pound tube of lead white. We never painted with pure color. Mother patiently explained how highlights and shadows fooled the eye, how any color could be made by mixing two or more tubes together. I vividly remember a particular portrait with a white shirt. We used white mixed with black, white mixed with yellow, and white mixed with blue to paint that white shirt. What a fascinating illusion for art to use everything but white to paint something white. Helping her mix colors on that palette was a treat for my senses. They were beautiful to look at, but their fragrance was an even greater delight.

I took her canvases down to the basement to dry. Oil and turpentine wafted upstairs, mingling with medicine and disinfectant. Only years later, recalling the incense of this conjugation, did I realize the influence that these two worlds had on me and how unusual such a childhood was. I take having one foot in science and the other in art as natural. Yet, as C. P. Snow pointed out, most people see an unbridgeable gulf between them. I learned much later that the gulf is an illusion. But I'm getting ahead of my story.

Another memorable smell came from my own basement workbench. It was the acrid fumes of electric motors and melting solder. I loved to take things apart to see what made them work. Other kids had sharp eyes for baseball gloves or model airplanes, but I was more likely to spot a screw head on a machine, an invitation to take the cover off and look inside. Having to know what the gears and vacuum tubes did drove me to dismember clocks, old sewing machines, record players—whatever I could manage.

From time to time I was more skillful at taking things apart than at putting them back together. But this failed to discourage me. In fact, it taught me a lesson that would be useful later, namely, that things are often more than the sum of their parts. In time I turned to electronics. There, the moving parts were not gears and belts but charged particles, the electrons of electricity. Capacitors and transistors did no more than shuttle these electrons from one place to another, but such electric switches could make pictures on a cathode ray tube or carry a voice across the ocean. This was the magic of television and radio.

Mesmerized as I was with wanting to know what made things work, it was small surprise that I later turned to neurology, a study of what makes humans work. Wanting to know how machines and electronic circuits worked as a youngster matured into questioning how the mind worked. How do we think? What drives our wants and desires? Why do we value some things and not others? How is it that we can have subjective views in an objective world?

Imagine what shot through my mind when Michael Watson said, "There aren't enough points on the chicken." Listening to his dismay at how the *taste* of roast chicken made him feel a round *shape* in his hands, as if he were rubbing a bowling ball instead of feeling the prickly shape he had expected, I felt the stirring of my take-it-apart instinct. I had to know what had apparently cross-wired his senses. My boyhood tinkering had taught me that what seemed complex could often be understood by looking at its parts. I believed I could find an explanation for Michael's strange perceptions, and that the answer would lie in his brain.

Most people at the end of the '70s would have just dismissed Michael's odd comment as flaky artist talk. Physicians at this time especially would have been quick to belittle human experience that did not conform to inflexible ideas of how things were supposed to be. Synesthesia is one of the most obscure medical conditions, affecting only ten in a million people. I later discovered that only two of us among two thousand people at the medical center had even heard of the term. I would find that the readiness of others to dismiss direct experience had little to do with the fact that synesthesia is rare. Recognizing that Michael's offbeat comment was a symptom of some arcane medical condition had less to do with my personal interests than with what medical education was like in 1973. Understanding that climate is important because those students are today's practitioners.

Meeting patients is not, surprisingly, part of early medical education. The first years of medical school were actually a question of whether the mind can absorb more than the seat can endure and of how well one can parrot back a professor's lecture. We either sat all day, feverishly copying every utterance, or else we were stationed in the lab dissecting the dead, peering through microscopes, or practicing surgery on rabbits (our team's died suddenly from the anesthesia). In theory, this activity was the foundation for understanding how the body worked, but with each passing month it seemed less relevant to what we imagined being a real doctor was like. A knowledge base is vital, but so is understanding real people, their needs, their feelings, and what their illnesses mean to them.

My optimism perked up in the second year when neuroscience would be the main subject. At last, I thought, here comes an explanation of the brain, the mind. Our last task in gross anatomy before summer vacation was to take the brains out of our cadavers and store them in Tupperware containers filled with preservative. Returning in the fall, I was eager to dissect them. What a disappointment that those brains soaked in formaldehyde for six more months while we soaked up more lectures and microscopic slides in daunting detail. I drew pictures of the nervous circuits to help remember where they went and what they did. Bur more were added each week that overlapped or frankly contradicted what we had already memorized. It was an avalanche of conflicting facts, with more exceptions than rules.

Just before final examinations a new lecturer appeared to spend the last three days of the semester on something called the "limbic system." We figured it was some minor add-on, coming so soon before the exam. What a miscalculation! Her twenty-eight-page handout illustrated "a few" of the "major points" of this limbic system, which turned out to be connected forwards and backwards with everything we had memorized so far. It was another incomprehensible torrent of fact, conjecture, and exceptions. This last straw actually led to a riot. The dean quieted the rabble after we stormed his office, but nothing could assuage the resentment we felt at this last-moment burden. I was disgusted with neuroscience. In all the lectures there was not one mention of the mind.

I suppose in retrospect my eagerness was like that of an art student bridling to paint portraits when he had not yet mastered perspective. Yet how disappointed I was that in all those months the details we had to memorize had only to do with reflexes, twitching toes, and feeling pin pricks. If this was "how the brain worked," I was mightily disappointed.

I hated neurology. What was the point of being forced to memorize so many contradictory facts? My anger doubled and trebled while studying for that last exam. Frustrated by the contradictions, I gave in to a full-fledged temper tantrum, proclaimed the whole topic incomprehensible, and flung my notes off the apartment balcony.

▲ ■ ●

Obviously, I retrieved them.

Several events precipitated a dramatic change in my third year. Most important was finally seeing patients. These clinical rotations were six-week tours through different departments, such as Medicine, Obstetrics, Psychiatry, or Surgery. As fate would have it, one of my first rotations was Neurology.

I kept expecting something horrible, like a giant spider, to descend on me. But the experience turned out not to be so bad. Real patients put the disembodied facts I had to call on in a human context. Neurology patients had complicated afflictions. The strangeness of those afflictions intrigued me; the fact that nothing was clear-cut challenged me. What was wrong had to be sorted out as a detective solves a mystery, weighing clues and discarding promising but false leads. When other specialties could not figure out what ailed their patients, they often consulted the Neurology Department as a last resort. While some students found this prestigious, what I really learned was that neurology is a method more than anything else. Other specialties relied on concordances of symptoms and diseases, but neurology's method was as familiar to me as taking a machine apart in the basement, and seeing how each piece fit into the whole in order to make the whole work.

I was no longer a sponge passively soaking up facts in a lecture hall, but a person actively concerned with the heartbreaking problems of other people. Beyond this, however, the real catalyst for change was a teacher. In all the doctor-patient encounters I had observed, Dr. William McKinney was the first physician I had ever seen sit down on the bed, take the patient's hand, and talk. Dr. McKinney spoke not as an omniscient authority but as one person to another. The residents complained that Dr. McKinney's patients were not very exciting and that they never had anything wrong with them.[1] They wanted sick patients that they could do things to, not a rare lesson in the art of medicine from a master who practiced it flawlessly. In such a skilled clinician's hands neurology was no longer a collection of contradictory facts, but a method, even an attitude. That attitude was one of respect—for the people, certainly—but also for the bigness of the enterprise. It was a revelation. So was the reaction of my comrades.

There was still an enormous amount of reading to be done. Long hours and endless patient rounds were followed by late evenings of study. On one of those evenings, I changed forever.

The traffic below my window had long since become a trickle. The occasional sound of tires on the winding road was the sole punctuation of the night stillness. Even the crickets were quiet, and my head had grown as heavy as the book I held. But study I must, even at two in the morning.

To wade through this difficult subject was surely one of the lesser circles of the Inferno. I read about a parade of human misery—paralysis, blindness, intractable pain, seizures, and stroke—but each description was dry and lifeless. Take notes, push on, turn the page. A blast of cold air from the open window perked me up and urged me on.

Suddenly I laughed. No, not a laugh. A hearty, involuntary guffaw. "That's fantastic! Utterly fascinating!" I gasped out loud. The heavy tome had suddenly come to life. What had riveted my attention and set me wide awake was aphasia, a disorder of the brain. Damage in a specific part of the brain could suddenly rob you of language. Oh, you could move your lips and make sounds, but meaningful speech was suddenly gone. Worse yet, you were unable to read or even understand anything being said. Destroy a tiny area of brain tissue . . . and meaning suddenly vanished. I thought so diabolical an idea must belong to science fiction. But no. Here it was in my medical textbook.

Not only words, but gesture, semantic and syntactic meaning, even the melody of speech all vanished from the mind in an instant. That was aphasia. What a horrible plight. If aphasia took away someone's symbolism, what happened to that person's humanity? Did it remain, or was it gone too? I had discovered something rich in philosophical implications.

A sheepish grin spread across my face. Was I the same person who had pronounced the subject incomprehensible, a subject as confused as the tangle of brain circuits we were forced to memorize? I was, undeniably. Some part of me—a part I would understand much later—had laughed out loud while studying in the wee hours of the morning what I proclaimed was gobbledygook. If neurology was in fact sublime, if it could cause such a reaction while I fought off sleep, then maybe I should reëxamine my bias. Maybe my first impression had been wrong.

My astonished laugh at the sheer cleverness of aphasia was so unexpected given my earlier disgust with the avalanche of neurological facts. Yet at this moment I marveled at the profound difference between moving one's lips to make sound and the ability to communicate and understand. Aphasia was not a fact like the hundred facts I had memorized before, but a concept. The focus of my understanding was now transferred to a higher realm whose center was a gray furrow inside my head.

I was gripped by a powerful feeling, as if I had been tapped on the shoulder by a ghost. The world suddenly had become an open space of new possibilities, as if I had stepped through a doorway from a cramped chamber. I did not know it then, but it was the beginning of my career in what is called the "higher functions" of the brain. Only a few months before, I had rejected neurology by throwing my notes off the balcony in disgust. But that night, a career in neurology had chosen me.

The higher functions like language and aphasia were just what I had been seeking. As the name implies, they have to do with the brain's psychological and intellectual abilities, the mystery of the mind. The higher functions included memory, thinking, perception of spatial relationships, mood and personality, and even such philosophically deep capacities as value, judgment, and volition. Indeed, the line where neurology ended and philosophy began was thin.

A prepared mind finds what it looks for, and one product of my new-found vigilance was a paper I wrote on aphasia affecting the French composer Maurice Ravel.[2] The New York Times music critic Harold Schonberg mentioned that Ravel was stricken with aphasia, which had somehow affected his ability to compose. This was all Schonberg said. I set off to find out how it had affected Ravel's mind, reading biographies and eventually tracking down Ravel's physician, the famous neurologist Théophile Alajouanine. He was still alive and living outside Paris. Because I was ignorant of his stature, I was bold enough not just to write to him, but to do so in imperfect French. The old master was generous enough to correspond with a lowly student.

Aphasia struck Ravel at the age of fifty-eight, quelling any further artistic output. Most horrible was the dissociation between his ability to conceive and his inability to create. That is, he continued to *think* musically but was unable to *express* that music by writing, playing, or singing. Rarely in aphasics is there an extraordinary development of the right hemisphere's skills as in Ravel. In most right-handed persons, musical ability depends more on the right hemisphere whereas language is exclusively a function of the left. Aphasia is usually a left-hemispheric problem. But in professional musicians, like Ravel, the situation is more complicated. For them, music is more than sounds and rhythm—it is a medium of communication, and *both* hemispheres contribute to musical expression. Like a writer who no longer can translate ideas into words, Ravel could no longer translate the right-hemispheric patterns which were his music into left-hemispheric symbols. As a result, his new music remained silently trapped by his aphasic mind and the world never got to hear it.

Exploring how Ravel was unable to write down or perform the new music that he composed in his head led me to dig into the new split-brain research and the differences between the two hemispheres. This body of knowledge was just emerging in the 1970s. It grew as a result of more frequent surgery for epileptics with uncontrollable seizures, an operation that cut all the connections between the two sides of the brain.

On the face of it, this is a drastic thing to do because more nerve fibers cross back and forth between the hemispheres than come in from the outside senses. Cutting these millions of fibers so that one side does not know what goes on in the other side should surely wreak havoc on physical ability as well as the mind. But in casual conversation, and even during a standard neurological examination, these people look and act normal! This is a hard situation to accept because it is so counter-intuitive. With special testing that restricts input to only one hemisphere, however, it is easy to show that splitting the brain does do something drastic. It reveals a wonderful paradox.

The "person" who speaks is not the person who perceives or solves problems. There are at least two separate personalities in each of us, but the unification of the hemispheres by the cerebral commissures gives the seamless illusion of a single, integrated self—namely, the person who speaks. In patients in whom the cerebral commissures have been cut and in whom experimental test input is ingeniously restricted to only one hemisphere at a time, the illusion of unity disappears. The right hand literally does not know what the left one is doing. One hemisphere can solve a problem for which it is well suited, while the other one responds with surprise, ignorant of what is going on. These hemispheric differences are not a product of the surgery but normally exist in everyone.

What split-brain patients show by this situation is that language is only *one* intellectual function. Because only humans can speak, we arrogantly assumed for years that language was our highest ability. It turns out that language is only *one* ability. Not everything we are capable of knowing and doing is accessible to or expressible in language. *This means that some of our personal knowledge is off limits even to our own inner thoughts!* Perhaps this is why humans are so often at odds with themselves, because there is more going on in our minds than we can ever consciously know.

At last I was uncovering the kind of sublime and complex situations that first attracted me to neurology, those where truth was stranger than fiction. The real truth that scientific digging uncovers is often counter-intuitive. This is the definition of a paradox: something apparently inconsistent with itself or with reason, though in fact true. I loved these odd facts, particularly those that contradicted long-held dogmas based on no higher authority than common sense, or what "everybody knows." I took as my motto that of the Royal Society, *Nullius in Verba*,[3] best translated as "take nobody's word for it; see for yourself."

Well, I did.

Chapter 4

▲ ■ ●

HOW THE BRAIN WORKS:
THE STANDARD VIEW

For those readers who watch television programs like "Nova" or read popular books about the brain, nothing I am about to say will be new, even though all of it is wrong.

Typical surveys found in popular books and programs, what one might call "Neurology 101," explain how the brain works in terms of an organization we conceived of at least twenty to thirty years ago. The three prime concepts of this "standard view" are that information flow is *linear*, physical and mental functions can be *localized* to discrete parts of the cortex, and that there is a *hierarchy* which makes the cortex supreme, dominating everything else below it.

The standard view is no longer standard, although it was very much alive at the time I first stumbled upon synesthesia. In fact, parts of it are somewhat useful for reasons I will explain at the end of this chapter. Even though the "standard view" has declined into the "old view," its three basic concepts are worth repeating because they are central to our medical mystery tale.

The flow of nervous impulses (information, if you like) is conceived of as linear in the same way that a conveyor belt goes through a factory. One piece is added on top of another until a finished product rolls off at the other end of the line. Sense impressions coming in and motor actions going out of the brain were conceived of in this way. I will only talk about sensation since this is what concerns us in regard to synesthesia. The first step is for the sense organs to transform either electromagnetic energy (vision), mechanical energy (hearing and touch), or chemical energy (taste and smell) into nervous impulses. These impulses then travel to different relays in the brainstem and thalamus, and from there to progressively more complex stations of the cortex where different aspects of the external stimulus are sequentially extracted from the stream of nervous impulses. These aspects are somehow assembled at the end of the line into a conscious experience so that we understand what it is in the external world that has triggered our sense organs.

Localization of function is the second major tenet of the standard view. For example, the occipital lobe is concerned with vision, the parietal lobe with touch, and the temporal lobe with hearing. The division of the brain into "lobes" was done so long ago as to have no actual validity. The several schemes for dividing the brain into forty or more physically discrete units are all based on the microscopic patterns of how cells are arranged (the technical term is cyto-architecture). Those who mapped the brain's cellular architecture at the turn of the century were surprised to discover that the discrete areas they had found looking through their microscopes did not at all follow the natural boundaries of the brain's bumps and fissures. Yet, the names of the four different "lobes" are still used as a general point of physical reference as Figure 1 shows.

The word *cortex*, which means "bark," refers to the bumpy surface of the brain. It is also called "gray matter" because of its color. The cortex is the largest of all brain components and has the most complicated architecture. It is also the youngest part in terms of evolution. For these reasons, together with the fact that the human cortex is far more developed than that of other animals, the old view understandably pointed to the cortex as the essential matter that distinguishes us from other creatures. In the history of trying to understand the brain, however, we have emphasized it nearly to the exclusion of everything underneath it, one practical reason being that it *was* the surface and could be easily approached experimentally.

The idea of a dominant cortex exists even in the triune brain model pictured in Figure 2, even though its originator did not intend it. The concept of three-brains-in-one was first proposed by Paul MacLean in 1949 to show that human brains contain three systems, each of different evolutionary age, and each governing a different category of behavior. The oldest is the reptilian brain, represented by brainstem and basal ganglia, and deals with self-preservation. The later paleo-mammalian brain is our inheritance from the mammal-

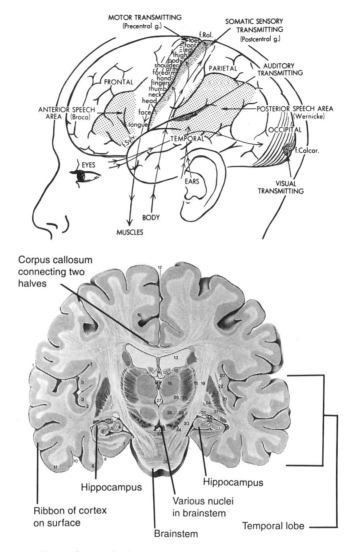

FIGURE 1. Top: *The standard view of how the brain is organized is a nineteenth-century concept that usually shows only its surface, or cortex. The gross "lobes" and general location of primary areas for vision, hearing, touch, and movement are indicated, as are the "motor" and "sensory" speech areas. Taste and smell are never shown, even though the standard view assumes that every function is represented in the cortex. From Popper & Eccles* The Self and Its Brain, *p. 229, 1977. New York: Springer Verlag.*

Bottom: *A cross-section through the front portion of the brain. Despite all the fuss over it, the actual ribbon of cortex (the perimeter shown in darker gray) is just 1 millimeter thick on average, a fraction of the brain's total bulk. The hippocampus is one of the most readily recognized parts of the limbic system, all of which is*

buried deep beneath the cortical mantle. Modified from Nieuwenhuys, Voogd, &
van Huijzen, 1985. The Human Central Nervous System, 3rd ed., p. 41. New
York: Springer Verlag.

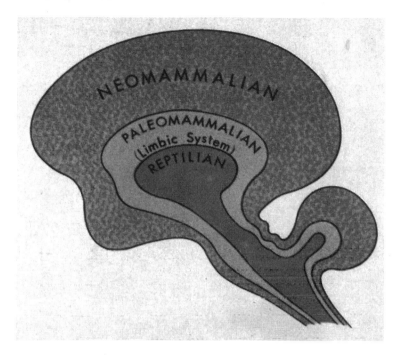

FIGURE 2. *The triune brain of Paul MacLean conceives of three-brains-in-one,*
each segment relating to particular categories of behavior and reflecting a distinctive
inheritance from earlier life forms. With permission from Paul MacLean.

like reptiles and is concerned with preservation of the species (sex,
procreation, and socialization), plus uniquely mammalian behaviors
such as nursing, maternal and paternal care, audiovocal communication, and play. Collectively, the components of the paleo-mammalian
brain are called the limbic system, which in humans represents the
emotional brain. The newest of our three brains, the neo-mammalian
brain, is represented by the huge expanse of cortex and is seen as a
chief executor that dominates the other components. The triune
brain's apparent agreement with the old view's concept of hierarchy
is evident from Figure 2.

In 1902, the British neurologist Lord Sherrington showed that the
brain's central fissure separates a "pre-central" motor area in front of
it, while in 1909 the American surgeon Harvey Cushing showed that

a "post-central" sensory area lies behind it. A central fissure dividing a motor brain in front from a sensory brain in back exists in all placental mammals. In 1952, electrical mapping showed that the organization of the sensory area behind the central fissure is a mirror image of the motor area in front of it. The existence of two spatially separated areas, each with its own function, became a fundamental concept of neuroscience. Detailed electrical mapping of the cortex done during surgery held out the hope that we could establish a point-to-point correspondence between brain tissue and function, both physical and mental! (Such materialistic reductionism has since been abandoned.)

Within the sensory half of the brain, the "primary" sensory areas were determined for vision, hearing, and touch. Those for taste and smell were debated for a very long time, but the point is that all sensation was believed to have a cortical representation. The "primary" areas were the first cortical relay station, damage to which caused total loss of function—such as blindness, deafness, and aphasia. "Secondary" association areas were soon found for each sense. These were conceived of as relays further along on the conveyor belt of perception, and which received more highly processed information. Damage to these secondary areas caused a distortion of that sense rather than its total loss. An example is the failure of recognition called *agnosia* (meaning "not knowing"). In visual agnosia, for example, you can see and describe an object, but neither recognize it nor know what it is used for. Agnosia can occur in any sense. At the end of the line was the "tertiary" association cortex in the parietal lobe. This was where sight, sound, and touch converged and where associations between and among the senses were made. While each sense had its own secondary association area, there was only a single tertiary area, where the highest, most abstract levels of reasoning were believed to take place.

You will notice that smell and taste do not fit this conceptual picture because their cortical representations are far removed from the tertiary association area. Instead of addressing this flaw, the standard view dismissed smell and taste as less important than sight, hearing, and touch. Another function that received only passing attention was emotion, a human trait known to be served in great part by structures beneath the cortex. If scientists considered emotion at all, they conceived of it as a detour branching off the linear information stream. Even then, emotional calculations were conceived of as secondary to those that took place in the cortex itself.

These three ideas yielded one more important concept, the assumption that *the cortex is the seat of reason and the mind, those things that make us human.* The accident of Phineas Gage in 1848 is the classic example of this idea. Gage, a railway foreman in Vermont, received history's first frontal lobotomy when a hefty tamping iron was propelled through his left eye socket, passed through both frontal lobes of his brain, and exited through the top of his skull. After a brief collapse and convulsion, Gage was up and about—albeit with a new personality. As his work mates put it, he was "no longer Gage." He was uncharacteristically compulsive, obstinate, vacillating, irreverent, vulgar, and lacking in self-restraint. He showed what we now consider typical signs of a frontal lobe disorder.

Beyond the fact that someone could survive such an injury at this time in medical history (Lord Lister, whose name we trivialize with a mouthwash, had yet to invent antisepsis), Gage's injury was astounding because it affected his "higher" mental functions, his personality rather than gross sensation or the ability to move his limbs. A small strip of the frontal lobe, the pre-central gyrus, had already been associated with movement, but its greater bulk could not be assigned any obvious function. For this reason the frontal lobes were called "silent areas" for many years because damage to them did not cause any obvious motor or sensory symptoms! This is a prime example of circular reasoning. Because Gage lived, his accident was concrete evidence that the brain really did have something to do with the mind. Today, the conceptual poverty of simplistically dividing brain function into only movement and sensation is obvious. Being human is much more than this. But even past World War II, neurologists did not think this way.

For very curious reasons, neurology was not at all interested in the mind during its early development as a scientific discipline. Indeed, up until 1950 many prominent neurologists firmly believed that the brain was concerned only with twitching toes and nothing more. They emphatically denied that it had anything to do with human behavior.[4] This attitude guaranteed that twentieth-century neurologists would reject synesthesia as way too subjective. Historically, general interest in synesthesia peaked between 1860 and 1920, then quickly faded because no one had succeeded in giving it a physical explanation. Even the standard view summarized here, which explains the linear division of sensory representation and its convergence in the tertiary association area, came too late to be useful to those who were interested in synesthesia during its heyday.

When I first sought to explain synesthesia many years ago, the standard view guided my thinking. It obviously suggested that the tertiary association area of the parietal lobe would be the seat of synesthesia because this is where at least three senses converge. The most obvious hypothesis was that there was an overlap of function, an entanglement of synapses, or that neural impulses belonging to one sense were somehow transferred to another. In other words, the most intuitive explanation was crossed wires.

An experiment to prove this idea would have to involve two essential steps. The first would yield objective evidence that synesthesia was real rather than made up or imagined by the people claiming to have it. The second step would show that the mechanism of synesthesia was located in the tertiary association area of the cortex. Hopefully, one could also show how the wires got tangled up.

As I was soon to find out, however, this line of thinking was entirely wrong because the standard view of how the brain is organized was wrong.

To summarize, the assumptions in the old view of how the human brain works are that it is linear and therefore something like a machine. The metaphoric likening of the brain, reason, and the mind to a machine is well known and extensively written about. The concept of hierarchy makes the cortex the brain's most important part. This part of the standard view says that the cortex is where consciousness, mind, reason, and reality are all located, and that everything below it is literally subservient. An important corollary says that language is the supreme cortical function; therefore, introspection, which is our self-conscious internal talking to ourselves, is a valid way to understand everything that goes on in our minds. Introspection has a long history in the philosophy of mind, but I will show its severe limitations and that we actually have several concurrent streams-of-consciousness running every moment. That not all of them are accessible to language has important implications for what we can "know" in the conventional sense.

Shelves of books have been written on these ideas. My intention is not to rehash them, but to convey the conceptual mind set prominent in my student days and one still with the average scientist today, not to mention the general public. Perhaps it persists in part because it is easy to grasp and useful to a point—just as Newton's mechanics are still useful, even though everyone has known for decades that Einstein's relativity is a more accurate description of the universe. But the standard view also persists because no one has brought the public up to date regarding our changing concepts of how the brain works and laid out the big picture. Gleaning a detail here and there from television or print guarantees in a way that those who are interested never see the forest for the trees. You should be able to behold that new forest when we have concluded our adventure with synesthesia.

Chapter 5

▲ ■ ●

WINTERS 1977 AND 1978: "THERE IS NOTHING WRONG WITH YOUR EYES."

Dr. McKinney noticed my fascination with neurological problems and encouraged me. He even suggested that I had talent in neurology and urged me to make it my career. To show his support, he arranged a scholarship for me to study neurology at its English-speaking mecca, the National Hospital for Nervous Diseases in London, an institution better known by its address, Queen Square.

In London, the interaction between doctor and patient was valued more highly than tests. One example of this was the week-long waiting list to get a CT scan. This astonished me because the EMI scan (which is what CT scans were first called) was invented in Britain, and Queen Square was Britain's premiere neurological hospital. Back home in North Carolina, we routinely got same-day service for this brand-new technology.

Physicians in the United States loved tests and had an overwhelming indifference to higher mental functions. The British were just the opposite. In America, a patient complaining of forgetfulness would automatically get a CT scan, despite the fact that machines are incapable of measuring higher functions such as memory. In Britain, physicians spent time probing the history and amalgamating it with the results of their examination before they rendered an opinion. That opinion was their diagnosis, a word that literally means "through knowledge."

My time in London impressed upon me the limits of machines and the value of careful reflection. Dr. MacArdle, an elderly and very famous neurologist, was the guest professor for teaching rounds one afternoon. While matron served tea and cake, Dr. MacArdle listened to the chief resident present an unknown case. The professor's role was to ask questions, arrive at a diagnosis, and use the exercise to instruct the audience. The lesson he taught probably remains vivid for everyone who was in the room that cloudy day.

"This gentleman has been weak in his arms for about six weeks," began the resident. It was a case of peripheral neuropathy, an afflic- tion of the nerves in the extremities. He went through the exami- nation, demonstrating the man's lack of strength.

"Is it symmetrical?" interrupted Dr. MacArdle.

"Yes," answered the resident.

"So, what are you going to do next?" he asked. I had assumed that the discussion would focus on differential diagnosis, since the causes of peripheral weakness are many and difficult to pin down. Instead, the old gentleman chose to put the chief resident on the spot.

"He's scheduled for nerve conduction studies," answered the resi- dent, referring to electrical tests that would measure the speed of transmission in the nerve.

Dr. MacArdle set his tea cake down. "Bloody hell! Whatever for?"

"To show slowing," he answered, meaning that the test would show slowed-down conduction speeds consistent with the patient's weakness.

Dr. MacArdle paused to empty his tea cup. Then he simply asked, "What causes slowing?"

"Loss of myelin insulation around the nerve fiber," the resident answered promptly. "Less commonly, damage to the nerve fiber itself."

The professor faced us in the audience, pushed back a lapel, and hooked his thumb underneath his maroon braces. "Now," he asked, "surely we all know the clinical manifestation of slowing, don't we?" He quickly turned to the resident, who was fidgeting with his reflex hammer. "What is it?"

"Weakness," admitted the resident softly.

"And weak is what this man is," wheezed the old professor, his face turning pink and mottled. "So, do you have any doubt that the nerve conduction studies will be slow?" he asked. "What on earth do you want them for, then?"

"We need an objective record," suggested the resident, trying to defend his decision.

"You've got an objective record," exploded Dr. MacArdle. "He's weak. You've just examined him in front of all of us. Don't you trust your own eyes? Do you need a machine to make your decisions for you?"

They volleyed back and forth, the resident eventually failing to justify why he really needed a test. In Dr. MacArdle's book, tests were not to be used to restate or "document" the obvious, nor were they licenses to go fishing. He particularly hated the latter, in which you ordered everything you could think of and waited to see what came back positive. "Only the Americans do that," he scolded. Tests were to be gotten only when absolutely necessary, and then only to *confirm* a diagnosis you had already arrived at in your head.

▲ ■ ●

I returned to North Carolina, received my M.D. degree that spring, and remained to begin my residency in a subspecialty called neuro-ophthalmology. Simply put, it was a combination of neurology and ophthalmology that focused on visual disorders originating in the brain rather than in the eye. This made sense for would-be neurologists because vision is so often affected by neurological illness. In fact, the eye is actually part of the brain.

One of the things I noticed while training in ophthalmology was how often patients complained of funny distortions in their vision. They were bothered by things like colored spots, sparkles, haloes around lights, moving shadows, annoying lines, and distortions like heat waves. They mentioned these matters almost apologetically because the ophthalmology faculty was not interested in hearing about them.

"By the way, Doctor Fleming, what are those lines I see?"

"What lines?" the doctor fired back.

"It looks like a worm, like a curvy blob on my right side," Mrs. Bates started to explain, drawing an S-shape in the air with her white-gloved hand.

"A worm?" marveled the doctor, not stopping to look up from the chart he was writing in.

"Of course, it isn't a worm," she said carefully, "but it resembles one." Dr. Fleming's pen scratched in the semidarkness. "When I read, it comes and goes. I rub my eyes but it stays. Later, I notice that it's gone."

"Your medical doctor asked me to check up on your glaucoma, Miz' Bates."

Mrs. Bates was a proud, Southern woman whom one could aptly call genteel. "Yes, Dr. Fleming, I am glad it has improved," she agreed with meticulous diction. "But I have worried about this worm-like spot."

Dr. Fleming snapped the chart shut. "I suppose it could be a floater."

Her voice rose slightly in pitch. "A floater? Do I have floaters?"

"I didn't see any."

The possibility of a floater suddenly animated her. "I certainly don't want floaters," she exclaimed. "I do like to read so." She gathered herself a moment. "Exactly what *is* a floater?"

"It's a piece that breaks off from the inside and floats across the retina," mumbled the doctor as he swung the phoropter in front of her and turned its lens wheel. Its clicking sounded like the tumblers of a safe.

"If it isn't a floater, Dr. Fleming, then what is causing me to see that worm shadow?" she asked again.

"Look at the blue light on the far wall, please," answered the doctor as he blasted a painfully bright light through her dilated pupils. She decided not to tell Dr. Fleming that he was blocking her view of the blue light on the far wall. She looked at his ear instead.

After a few "hmmm's" to himself, Dr. Fleming spoke at last. "I don't see anything wrong with your eyes except for the pressure being mildly elevated."

"And the shadow?"

"We need to talk about your glaucoma medicine."

▲ ■ ●

Mrs. Bates' worms and other patients' what-nots were subjective complaints. This means that while they "saw" something, the doctor looking into their eyes couldn't "see anything." Since the faculty could not see it or measure it with a machine, they decided that whatever patients were seeing was either unimportant or not real.

As expected in any large medical center, patients were referred because they usually had a problem in a given faculty member's specialty. The faculty focused on cataracts, retinal degenerations, or whatever their field was. They only looked for what they wanted to see, dismissing other problems out of hand or at best explaining them away as "floaters."

I looked into these people's eyes and was unable to see anything either. But I was struck by how exact and detailed the descriptions were of what they claimed to see. It is true that floaters are real in middle-aged people and can occasionally cause them to see moving black blobs. But surely floaters could not be the explanation for the dozens of things patients saw. After hearing a goodly number of precise descriptions, it seemed that they fit patterns, patterns that sounded too much alike for hundreds of patients to be making them up. Besides, floaters are objective. They *can* be seen. So why blame floaters at all in cases where we could not "see anything"? There had to be a better explanation.

There was. The answers lie in dusty textbooks. I guess trainees were so busy learning to use lasers and other latest instruments that no one had time to read the actual words of revered authorities whose names were uttered daily. I read that some of these visual illusions were caused by problems with the eye itself. Swelling of the cornea could cause rainbows and haloes around lights, while retinal problems could cause flashing lights or colored arches. But the most unusual experiences were caused by the visual parts of the brain itself. Some conditions were astounding. They would have been unbelievable had they not been recorded in texts by the grand elders of medicine.

Visual agnosia ("not knowing") is a condition in which the patient cannot recognize what he is seeing. This problem was made familiar in *The Man Who Mistook His Wife for a Hat*. Reversed and inverted vision, known by their German names of *Umkehrtsehen* and *Verkehrtsehen*, are astounding reversals in which the patient's world looks as it would if he were standing on his head! (Even stranger, patients are not bothered by this bizarre rearrangement.) Autoscopic vision ("self seeing") is an out-of-body experience. One woman was walking to work when she suddenly felt odd and then saw herself treading down the sidewalk. She watched for several minutes until her "other self" vanished. Similar episodes, such as floating up to the ceiling and observing one's action in the larger scene, are perhaps more familiar because they have also been written about as part of near-death experiences.

Other examples are polyopia ("multiple seeing") and palinopia ("old seeing"). One patient with polyopia looked at a single rose in a bud vase, turned away, and saw a bouquet of roses projected on the wall. Palinopia is like instant replay. A man with palinopia watched his wife leave his hospital room. A few minutes after she was gone, he saw her leave again. Metamorphopsia means "change-form-sight." It is the distortion of shape and size. Stationary objects seem to grow larger or smaller, as in *Alice in Wonderland*, or the scene will suddenly break into pieces which slide over one another as in a cubist painting.

I read that brain damage can cause hallucinations of colors and elementary visual perceptions such as sparks, flames, or flickering. The visual scene can take on a monochromatic hue, as if a color wash were painted over everything. Sometimes color even takes on a life of its own and refuses to stick to the boundaries of objects. In this case, color melts off an object as in a Dali painting. When one patient touched something, his finger seemed to sink into the color.

These visual errors amount to hallucinations in sane people. All are caused by illness in the brain. These forgotten but well-described examples certainly would explain a lot of "things" that patients were seeing. When I mentioned them to my fellow residents and the faculty, I was told, "That's interesting, but these things must be pretty rare because I've never seen them."

But they had. Or at least they would have if they had paid attention to what patients were telling them. They were looking but not seeing, and when you are only interested in cataracts, that is all you will see. I could not blame them for not being interested in these cerebral distortions of vision because they were ophthalmologists, not neurologists. But it did bother me that they were looking at the eye as if it were all that mattered. The eye is the first stage of vision and is located in the front of the head, while the visual cortex is in the back of the head. There was an awful lot of important stuff in between that they never bothered to consider.

I also sensed that a prevalent attitude, no matter what the specialty, was that the history of medicine had nothing to teach the present, and if symptoms could not be measured with a machine, then they were imaginary. All around me I found people willing to trade in their own judgments for ones made by a machine. Anything from the past was thrown without question on the scrap heap with the leeches.

▲ ■ ●

If you want to learn new things, you should try reading old books. I found this aphorism wise, especially given the information overload found in any technical field such as medicine. In 1976, I had already encountered statistics that the average practicing physician read eleven journals a month in an effort to keep up. Medical science was changing so fast that half of what new graduates knew would be obsolete in five years.

Hospital life was chaos, with constant requests for your attention pulling you in several directions at once. Even Avalokiteshwara, the Buddha with dozens of arms, would have been worn out by an atmosphere that demanded that everything be treated as equally urgent. The din of simultaneous conversations and the page operator's stream of announcements provided a background drone through which one heard the screech of beepers, the shouts and moans of patients, the clatter of bedpans and meal trays, and the roar of the buffers that the janitors constantly plied on the white linoleum. It was a miracle that sick people got any rest.

I took my own rest breaks in the library stacks, a quiet eye in the storm that surrounded me. Since people rarely went to the sub-basement levels, these held my favorite nooks. Sometimes I would just sit in this haven of tranquility for ten minutes, somehow hoping that a private, inviolate moment would maintain my sanity. At other times, I would browse among the books, noting what I wanted to put on my reading list. Reading books is an antidote to urgency and chaos because it cannot be rushed. Every book has its own rhythm and a physical intimacy that E-mail and similar instant information can never achieve. Media glut often confuses information with understanding. Just when you seem to be most pressed, books miraculously expand time for reflection, cogitation, and mental rest.

One of the books that had quickly made it to my reading list was called *The Mind of a Mnemonist* by the Russian psychologist A. R. Luria.[5] He called it "a small book about a vast memory," which it was. Luria had studied his patient, "S," for thirty years, trying to understand S's indelible memory. His subject was really S. V. Shereshevski, who had refined his talent to become a professional memory expert. Even more fascinating than Luria's account of S's photographic memory was the description of S's synesthesia.

Luria described both phenomena in detail, although he ultimately never offered a possible explanation of either S's prodigious memory or his synesthesia. S was not aware of any distinct line separating vision from hearing, or hearing from any other sense. He could not suppress the translation of sounds into shape, taste, touch, color, and movement.

Presented with a tone pitched at 2,000 cycles per second, S said, "It looks something like fireworks tinged with a pink-red hue. The strip of color feels rough and unpleasant, and it has an ugly taste— rather like a briny pickle . . . you could hurt your hand on this."

This same synesthesia enabled him to visualize vividly each word or sound that he heard, whether in his own tongue or in a language unintelligible to himself. The thing to be remembered automatically *converted itself* without effort on his part into a visual image of such durability that he could remember it years after the initial encounter. So specific was his ability that the same stimuli would produce the exact synesthetic response.

S was a person who "saw" everything, who had to feel a telephone number on the tip of his tongue before he could remember it. He could not understand anything unless an impression of it leaked through all his senses. Here is how he described the strange world in which he lived:

> I recognize a word not only by the images it evokes, but by a whole complex of feelings that image arouses. It's hard to express . . . it's not a matter of vision or hearing but some over-all sense I get. Usually I experience a word's taste and weight, and I don't have to make an effort to remember it—the word seems to recall itself. But it's difficult to describe. What I sense is something oily slipping through my hand . . . or I'm aware of a slight tickling in my left hand caused by a mass of tiny, light-weight points. When this happens, I simply remember, without having to make the attempt.

This was my first encounter with the term "synesthesia." It was because of Luria's book that I deduced what Michael Watson was talking about when he referred to points on the chicken. I found the term again in the dusty ophthalmology textbooks that dealt with subjective visual experiences, such as polyopia and *Umkehrtsehen*, as well as the worms, arches, and other "things" that patients ventured to acknowledge. Only one other reader had checked Luria's book out of our medical library, and few of my fellow residents even cared to hear about it.

"That's pretty weird," said Marty, one of my colleagues. We were preparing for rounds in our joint residents' office. "Sounds like too much dope."

"It's in *Walshe and Hoyt*," I countered, appealing to the authority of a standard textbook.

"It's got to be hysterical," another voice chimed in.

"Why does it have to be hysterical?" I asked. "Nobody who has written about it over a hundred-year period thought it was hysterical."

"All right, they're psycho, then. It's a hallucination, like hearing voices," Marty insisted. "I just can't believe that a normal person is going to see things that aren't there."

"They are not psychotic . . ."

"How do you know?" interrupted Mark, the chief resident. "You must have to be pretty screwed up to hear colors. Maybe these synesthete people are psycho retards, with their brains all scrambled."

"What about a temporal lobe seizure?" offered Justin, one of the neurology residents. "It sounds like a seizure. Seizures can make you experience some weird stuff, man."

"Sorry, no seizures in anything I've read," I answered.

"What physical signs do they have?" asked another resident.

"None."

"Oh, come on, they have to have a lesion, man! Signs and symptoms mean a hole in the head. It can't be real without physical signs."

"I can't say," I said. "The cases are forty, sixty, maybe a hundred years old. No CT scans, you know? This was when doctors would just write about what they knew."

Justin laughed and shook his head. "You can't rely on junk from old books, Rick. That stuff is all out of date."

"Why?" I asked.

"Because things change. Look, this is a stroke center. Some gork comes in all stroked out. You look for hard, physical signs and then work it up to find the lesion. That means an angiogram, a CT scan, or an autopsy."

"That's the way it is," agreed Mark. "You have to have proof of a lesion and the lesion has got to match the symptoms."

"Anatomy hasn't changed in five centuries," I said. "At least that's how long we have had drawings of it. Have we evolved into something physically different since Vesalius and Leonardo did their dissections?"

"Anatomy is different," said Marty.

"Why is it different?" I asked. "Function follows structure. If anatomy doesn't change, neither should perception. Why can anatomy stay the same, but perception can't?"

"Because it's just different," insisted Marty. "It just doesn't sound scientific."

"He's right," said Justin. "If synesthesia is for real, then why don't we admit people who have it to the ward? Why don't we all know about it?"

"I assume it's not very important," I said, "at least in terms of not causing any medical problems."

Marty interrupted. "Some slob hears colors and that's not a problem? It sounds like a major problem to me. Forget it, this is not on your neuro hit parade because it's not real."

"They're messed up, I told you," said Mark. "Come on, we gotta get rounds started."

"Maybe you haven't seen it because your mind is closed tighter than your sphincter," I told Justin. "You have to draw this kind of experience out of people," I said. "If folks are going to get this kind of hostile reaction from their doctor, who would ever talk about it?"

"I got enough work to do without dealing with crazy people," Mark said. "Admit'm, work'm up, send'm home, and roll in the next one. I don't want to hear about crazy chickens from people whose brains are scrambled. Somebody says a lot of crazy stuff to me and I'll pump him full of Thorazine and lock him up. Take my advice, Rick, leave this kind of stuff to philosophers. They have time for meaningless digressions. We've got work to do."

Chapter 6

DIRECT EXPERIENCE, TECHNOLOGY, AND INNER KNOWLEDGE

I would have assumed that the charming strangeness of synesthesia and the peculiar visual conditions I had uncovered would have made knowledge of their existence more common. After all, neurology is full of strange, unexpected, and even incredible clinical facts. Yet I found just the opposite. Synesthesia's respectable two-hundred-year history in the annals of medicine and psychology had been virtually forgotten. Even odder, I thought, was the hostility and doubt it evoked in those to whom I had mentioned it so far. I sensed from their vehement overreaction to these subjective experiences that my fellow residents somehow felt *threatened*. After all, could their view of reality hang on whether synesthesia really existed or not? I promised myself to get to the bottom of this fishy attitude.

The history of medicine teaches that certain diseases go in and out of fashion. Hysteria and swooning, for example, were rampant in the nineteenth century, but one hardly expects well-bred ladies to faint in indelicate situations today. Beyond a handful of cultural curiosities such as these, however, people's behavior—and here I mean the physiology of their perceptions—does not change. Human individuals do not evolve into Martians who suddenly have synesthesia. Since human physiology does not change, I reasoned, then it must be we, the observers, who have changed. And the reason we had changed is because we no longer observed human physiology directly, but through the lens of technology. Hands-on medicine had become passé.

Mark, the chief resident, had made this perfectly clear when I thought back on it. Even though he talked like a redneck, he was an intelligent man who had spent nearly twenty years educating himself to be a physician. Yet what he was most comfortable with was a black-and-white choice given to him by a machine. If the CT scan could not show Mark that there was a hole in the patient's head, then nothing was wrong with that patient. Ambiguity, uncertainty, subtlety, and nuance were alien to his thinking. His narrowness was precisely what Dr. MacArdle had warned against.

Mark's stance was not a reaction to being overworked, nor of any specific circumstance. Rather, it reflected how medical science had already become too machine-oriented and not enough oriented toward people. Fundamentals and the art of medicine seemed to recede further and further from our attention. By the late 1970s, the National Institutes of Health (NIH) and the biomedical industrial complex had been growing steadily for two decades. Technology was king.

Who stopped to ask which of today's techniques would be tomorrow's leeches? Who even asked if technological gain in one place might not force a loss somewhere else? Who asked if the price was worth it? I would not deny that the science of medicine benefited from this technological explosion, but, like most blessings, it was mixed. The uncritical acceptance of each new marvel as being all for the best disturbed me, and surely the gains were overstated in absolute terms.

There is no doubt that our health has improved mightily during the past century, but most analyses agree that this resulted from better sanitation, hygiene, and childhood vaccination. Even antibiotics contributed little to the overall returns in life expectancy and reduced disability compared to these simpler measures. Yet we are addicted to technology and favor crisis intervention more than prevention. Because we put great resources into making newer and newer devices, not only do we intervene more but also more expensively. For example, the bulk of federal Medicare insurance for the elderly is spent keeping people alive in their last six months, trying to prevent what cannot be prevented. That many recipients of this intervention do not judge the quality of their life in those last months to be satisfactory is a dilemma for which we have no solution.

When I was a boy, I sometimes went with my father on house

calls. I got cookies, lemonade, and other pleasant diversions while Father took care of the sick. I may have had a firsthand view of more sickrooms than others my age, but early baby boomers whom I have asked can recall that a sickroom was once commonplace. They also remember, as I do, when people died at home, surrounded by family. How comforting is death today when one is isolated in the intensive care unit and cradled by machines?

We have paid with dollars and our humanity ever since the stethoscope appeared as the first instrument to come between patient and physician. The art of medicine has steadily yielded to the calculus of objectivity and the tabulation of hard data. This economy has inflation, too. Machine interposition has increased exponentially, until today we have hardly any touching and little real human contact. Patients have been reduced to objects, and physicians to dispassionate feeders of the machines.

As a group, physicians are becoming as evasive as politicians. A direct answer is hard to get because machines are used not as tools, as *extensions* of our minds and senses, but as *replacements* for them. One is told, "Let's wait until the tests come back." Machines have corrupted the word *diagnosis*, "through knowledge," which once referred to the physician's knowledge of the fabric of the human body as well as its spirit. Now "diagnosis" has come to mean a deferral to the machine. We seem to have forgotten that factual knowledge and technical competence are not the foundations of patient satisfaction in many situations.

Lest you think that my attitude was colored by youthful idealism, let me recount a bizarre incident that took place when I was later engaged in private practice. I had completed a Fellowship in neuropsychology and left North Carolina to take a position as chief resident in neurology at George Washington University in Washington, D.C. I was surprised to learn that, in 1980, a neuropsychological evaluation was unobtainable at any of the *five* teaching hospitals in the nation's capital. The discipline of neuropsychology was itself a century old, and the NIH began accredited training programs nationwide in 1973 to markedly increase the number of practitioners.[6] Yet when I arrived in Washington, as far as I could tell I was the only neurologist outside the NIH itself with competence in neuropsychological evaluation. I planned to offer this service when I entered private practice a year later.

This turned out to be nearly impossible. Other neurologists, psychiatrists, and even the courts requested my service. But getting paid for it was a bureaucratic nightmare. Insurers such as Blue Shield have a long history of reimbursing surgeons for procedures, or "doing something," while placing little or no value on clinical reasoning, the very heart of diagnosis, treatment, and primary care. I have no doubt that decades of this policy hastened our current overemphasis on technology. Since insurers insisted that "doing something" rather than thinking constituted the practice of medicine, physicians ordered tests so they could get paid.[7] In my case, it took nearly two years of education and diplomacy to get Blue Shield to authorize a neuropsychological evaluation as a "new procedure." Academic experts and the Medical Society supported me in the Kafkaesque absurdity of convincing them of its usefulness in medical practice.

What troubled Blue Shield was that neuropsychology used paper and pencil, and a few small hand tools rather than a huge machine with flashing lights to test memory, reasoning, visual-spatial perception, language, and other aspects of higher brain function. That it was valid and accurate in localizing brain lesions and diagnosing their cause, that international professional societies existed devoted solely to neuropsychology, or even that it had a rich and long history in the annals of neurology did not matter to them. Neither did its direct bearing in deciding whether treatment for a given condition was possible or a waste of effort and expense. First, bureaucrats rejected anything with "psyche-" or "psycho-" in its name as being irrelevant to medicine. Secondly, they could not understand how my sitting down for two hours in direct interaction with a patient, and then synthesizing the results with the clinical situation and knowledge of brain function could be a procedure. To them, "thinking" and "reasoning" did not equal "doing something."

In the sense that third-party insurers are bureaucrats, they constitute yet another "machine" that stands between doctor and patient. Bean counters with hearts of stone have replaced compassion and caring. Medicare rules that prematurely force some elders out of the hospital, and the current zeal for "managed care" are examples in which bureaucrats usurp decision making and base it solely on the inhuman *technical* criteria of cost and efficiency. In contrast with my effort on behalf of neuropsychology, I had no difficulty whatsoever explaining anything or getting approval to install a CT scanner in my office. In fact, everyone was thrilled.

We have long believed as a society that technology serves us—we believe it saves lives, makes our work easier, improves communication, and is mostly good. But I believe that, hardly realizing it, we have come to serve technology even though we intended for it to serve us. The machine is held in such high esteem that, in medicine, many implicitly believe that caring is what is left for physicians to do when technical intervention has failed.

A good example is the treatment of James Brady, the White House press secretary who was shot in the head during the 1981 assassination attempt on President Reagan. While my participation in Mr. Brady's medical care at George Washington University was marginal, I did write a cover story for the *New York Times Magazine* examining the public reaction to his brain damage.[8] Unlike Phineas Gage's injury, which shocked observers because it affected his mind instead of his body, Brady's injury evoked a peculiar, opposite reaction.

To a person, every commentator focused on his paralysis. The public reports from his surgeon also emphasized Brady's physical condition and suggested that he would return to work using a cane. His colleagues from the press hardly ever questioned whether his thinking was affected, emphasizing instead how he had retained his well-known sense of humor. Everyone took this as a favorable sign, even though his humor was more marked than before and sometimes got out of hand. His fellow reporters could not be expected to know that this state of exaggerated humor is called *Witzelsucht*, a German medical term meaning "a pathological compulsion to make jokes," often shallow and facetious. It is a classic sign of frontal lobe damage, just like Phineas Gage's new personality. Those who have seen Mr. Brady on television or at public events cannot now fail to see his impulsive interruptions to make puns and other jokes.

What did technical intervention do for James Brady? Most of our experience treating brain damage originally came from doctoring those injured in war. Our technical skills improved with each war, as did the ability to evacuate soldiers to field hospitals rapidly. Our impressive technical accomplishments saved many people who would have otherwise died. Everyone thought that this was noble progress until those very technical achievements made us suddenly question the quality of those lives we were now saving.

Like so many technical solutions, the results of this "medical triumph" ultimately forced us to confront the humanistic issues that society at large had avoided for decades. The world cheered when surgical technique saved Brady; only much later did people acknowledge that he is now living a qualitatively much different life. He is surely testimony to how the human spirit can face steep obstacles, but his injury just a decade ago also illustrates our focus on faith in technical solutions.

This deification of the machine is, of course, not unique to medicine or even to science. Many authors have written eloquently on this point. While constructed by humans, a technological society ironically *dehumanizes* us by transforming it faster than the human psyche can cope. Up to the time of the industrial revolution, the tempo of living was tolerable and the conditions of life remained reasonably constant from generation to generation. As a result, individuals could develop a sense of connection, which itself was often the foundation for a rich spiritual life. Change was gradual enough that each person had a mental snapshot by which he or she knew what to expect in life.

But today technology changes the conditions of life so rapidly that just as you get yourself related to one set of conditions, another comes along. This causes anxiety and displacement. It is impossible to develop stability and psychological depth in a rapidly changing world. By removing us from our human center I believe that machines have taken us away from the depths at which we really live and abandoned us to a superficial existence.

▲ ■ ●

The authoritarian message that "we are the experts and know how things are supposed to be" seemed pretty arrogant, akin to claiming a crystal ball through which the faculty always had an answer. It was found in all divisions of medicine. The ophthalmologists knew what you were supposed to see and not supposed to see, the psychiatrists knew if your feelings were normal or crazy, and the internists knew what aches and pains were real and which ones were imaginary. Their crystal balls told them that they were always right.

What physicians of this era actually meant, of course, was that they could not find any objective data to correlate with patient experiences. They sometimes were at a loss to explain why people felt the way they did. Having learned nothing but their objective paradigms, how could they possibly deal with exceptional encounters except to deny them and defend their position?

I doubt there was any malice in such doctor-patient encounters.

Yet a harmful miscommunication resulted nonetheless. Maybe doctors should have said, "I don't understand," or "I don't know." Unfortunately, the message that patients perceived was, "You are lying," "Your body is deceiving you," or "You are crazy." Only a few were strong enough to reassure themselves that their bodies were right. They could say, "This doctor is not listening to me." But most people yielded to the white coat. "The doctor knows best," they said, "and therefore I must be wrong."

Physicians then may have been arrogant, but I do not think we were malicious because we too were a product of a society that had increasingly devalued the individual. We all reflect our culture. Historically, the West has emphasized institutions and feared individual gnostic experience. By gnostic[9] I mean "having to do with inner knowledge," a kind of knowing that transcends classification and is beyond the limits of ordinary experience. This is a crucial point to which I will return, namely, that we do *understand* what we know intuitively even though it may be impossible to *express* what we know.

Institutions stand in contrast to the individual and to gnostic experience. They are products of human *civilization*, whereas inner knowledge is a product of human *experience*. Institutions, including medicine, science, and religion, are more interested in preserving themselves as institutions *per se* than in serving, let alone understanding, an individual life. They are self-interested and stress conformity to the institution's values. Conformity certainly describes the atmosphere of medical education of the recent past, and some would say it still does. It is true that, finally, increasing numbers of physicians now embrace more holistic approaches and appreciate that individual differences do exist and can sometimes be important. The best of today's physicians combine scientific methods with humanistic attitudes. But subjective experience is still very hard for most of us to deal with.

Because most trainees of my time accepted what they were told with little questioning, stock phrases like "there is nothing wrong with you" became part of physicians' lexicon across the country. Authoritarian phrases like these were like another handy tool to pull out of the black bag when you were not sure what was going on. Society had trained legions of biomedical practitioners who could expertly treat the body as if it were a broken machine, using sets of black-and-white rules, but who were poorly prepared to deal with gray zones.

Obviously, there are situations in which patients are dead wrong. Psychosis and hypochrondria are examples. But patients whose crime is not fitting the mold of objectivity are an entirely different challenge. A thoughtful person would have noticed not only the gap between medical objectivity and individual subjectivity but how large it was, how persistent, and how recurrent throughout medicine's specialties. A thoughtful person would have concluded not that a sizeable minority of patients were completely crazy, but that our own attitudes were wrong because they kept us from listening.

A direct challenge to this assumption would immediately provoke proof of its correctness: an emotional reaction of denial, resistance, anger, and an appeal to the doctor's authority and the way things "should be" strong enough to convince such a thoughtful person that he had hit the nail on the head. If patients got this kind of attitude from the attending physicians, imagine the rigid, paternalistic ambience in which trainees worked.

Others may have been wary of rocking the boat or of venturing into unfamiliar territory. Like my colleagues, I too had worked up plenty of patients with headaches, strokes, seizures, and spells. With time, doing the same things over and over had become a variation on a theme. Diagnosis is not that challenging when it is an exercise in refinement. Synesthesia was the challenging exception to the rule. My idealism and occasional disillusionment with the system were no secret to anybody. It would have been easier to accept things the way they were. But it was not my nature.

I was intellectually attracted by the complexity of the life of the mind, but I was disappointed, too, that any promise of an explanation was an illusion. No matter how many questions you answer, you are always left asking more. There is no such thing as final understanding because understanding is an endless process. Answering one round of questions only takes you to a higher plane of understanding that makes you ask a higher level of questions. The experience of living itself is such a process.

▲ ■ ●

"Adolescence is difficult enough without having painful differences mark you from the other kids," Michael said. He told me about an incident from his youth.

"One summer, I went to Indiana to attend a summer school for science at Purdue. I passed a maple tree and remember being grabbed by the most incredible smell. It was musty, very complex. The smell was like a piece of sculpture I could feel pieces of, it had such depth, such an incredible texture. I dragged others to the tree and stood there, gulping it in, saying, 'Can't you smell it? Isn't it great?' "

"What are you talking about?" asked one of his classmates.

"Trees don't smell," shot back another.

He blushed, hardly knowing what to say. "Of course it does," he stammered. "Smell it," he said, drawing in a deep breath. "It's wonderful."

"Don't be silly. There's nothing there."

"But it's so intense," shouted Michael. "I can even feel it, it's so intense." He started to tell them about his touch sensations, but quickly thought better of it. Other kids had made fun of him in the past when he told them. His face fell, knowing that there was no use trying to explain.

"What a basket case!" teased one of the boys. "Come on, we've got to get to chemistry class."

From that moment, Michael knew that he lived in a different world than others did. It was a lesson that others like him quickly learned.

▲ ■ ●

My attraction to subjective experience eventually led me to conduct extensive experiments with Michael. At first they were simple paper and pencil assessments that I was used to giving in neuropsychology. But in time I did use state-of-the-art machines to measure brain metabolism and other functions. Michael eventually endured drugs, radioactive gas, wires pasted all over his head, and catheters in his arteries. When our experiments reached this advanced stage, he became frightened by the possibility that he had gone too far. More frightening than the risk of finding a brain tumor or other serious abnormality was the chance that someone might point at him and say, "This is ridiculous. What are you doing these stupid experiments for?"

"I had lived with synesthesia all my life and yet I still doubted it was real," he later told me. "I became afraid that the experiments would backfire and prove my worst fear: that my synesthesia wasn't real, that it was just Michael being silly."

Here was proof of how deeply the technology ethic is ingrained in our psyche—Michael's fear of being proven wrong by a machine, his implicit belief that it, not himself, knew what was right and real, and his willingness to reject his own direct experience.

As I studied the problem of synesthesia, it dawned on me that all of us are pressured to reject direct experience. It is not just a problem faced by individuals whose experiences are somewhat unusual. I kept thinking about Mrs. Bates and hearing the statement, "There is nothing wrong with your eyes." The medical establishment kept telling people that their subjective experiences were not real.

"Why shouldn't they be real?" I asked myself. My fellow residents were not interested in even discussing the question. Then I thought of someone who might listen.

Chapter 7

▲ ■ ●

MARCH 25, 1980:
BLINDING RED JAGGERS

"You're kidding me," said Dr. Wood. He turned back from the window and blew out a stream of smoke. "What about smell?" he asked. "Does smell make your friend feel things in his hands?"

"I don't know," I shrugged. "I never thought to ask. It was a dinner party and I couldn't get into it with the other people there. He was embarrassed that I knew about his chicken and the points." Dr. Wood chuckled, two bursts of smoke shooting from his nose. "You should have seen his face," I said.

I had told the incident about Michael Watson to my section chief, Dr. Frank Wood, who was now supervising my fellowship in neuropsychology. Instead of the disbelief and dismissal that I had gotten from others whom I had told, Dr. Wood actually knew what synesthesia was. Even more surprising, he had read Luria's book and was listening to my story with interest.

He sat back at his desk, so cluttered with folders and papers that the surface hadn't seen light for years. He picked up his half-finished tuna melt. "It would be interesting to do some experiments," Dr. Wood suggested, taking a big bite. "You know, see if you could manipulate what he feels by systematically altering the stimulus." Even with a full mouth he managed to drain his can of Mountain Dew. Dr. Wood had a reputation for devising clever experiments, but he made a greater impression on me by his ability to think, eat, smoke, talk, and drink at the same time. It was a stunning act of motor coördination. "Would he go for it?"

"He's an artist," I said. "I don't know."

Both of us thought through different possibilities and objections, whether we could make something scientifically interesting out of my serendipitous find. At least the two of us could have an intelligent discussion about it, even if everybody else thought I was wasting my time.

"Too bad you only have the one case," Dr. Wood said after a while, leaning back and lighting up a new Winston. "It would be nice if you could compare him to others. Luria's case, for instance." He huffed on the match and threw it in the trash, overflowing with computer printouts and the wrappings of recent lunches. "But that was different, wasn't it? He saw things, didn't he?"

"Luria's guy had a couple of his senses joined together," I corrected. "His case may not be comparable to Michael's. You don't know how it works, do you?" I asked.

Dr. Wood shook his head. "Hell if I know. Luria didn't speculate?"

"Afraid not," I said. "The book was description only. Besides, Luria was more interested in documenting the photographic memory than the synesthesia."

Beyond having read Luria's book, it turned out that Dr. Wood knew no more about synesthesia than I did. We would both probably have let the matter rest, content just to have recognized the diagnosis in Michael, had it not been for an unbelievable coincidence that occurred two weeks later.

I stumbled across my second synesthete right in Dr. Wood's office. We were talking about biofeedback when my beeper went off. Victoria, a staff member, responded to its three shrill tones by clutching her head.

"Oh, those blinding red jaggers! Turn that thing off," she snapped at me.

I silenced the device immediately. "I know, the damn thing's so shrill it even drives me crazy," I apologized. I noticed that Victoria kept rubbing the left side of her forehead and waving her hand in front of her eyes.

"What did you mean by 'blinding red jaggers'?" I asked her.

"The lightning bolts," she said, as if it should have been obvious to us.

"What are you talking about?" I asked.

"Your beeper made me see three red lightning bolts, brilliant red, going up to the left." She kept rubbing her head. "It's usually not that strong, but it's given me a splitting pain," she continued. "It must be the pitch. So high."

Dr. Wood and I looked at each other. He looked back at Victoria, who was still waving the red jaggers away. I wondered what they must look like.

"Victoria," he asked, "do you have synesthesia?"

"Sure," she said.

We were stunned. To find two people with such a rare condition would be unexpected. But to find two such people in the same town and within a few weeks of each other was incredible. Ironically, Victoria was herself a psychologist. She knew what synesthesia was and had had it all her life, but she had never talked about it because she knew it would damage her credibility at the medical school. "If people found out that I hear colors, they'd think I was out of my mind. Who would trust a therapist who saw things?" she asked us.

"What makes you see things?" I asked. "Give us some details."

"Sharp, shrill sounds always do it," she said, "like your beeper, or ambulance sirens, crashes, screeching tires." She groped for the right word. "Sudden sounds like that. Sometimes music will do it, if it's loud enough and high enough in pitch," she continued. "I once thought that shrillness had a lot to do with it. For example, my dog's bark never makes me see colors, but I once heard a Chihuahua who drove me crazy with the sound of white spikes. But that can't be the correct explanation, because words and names sometimes have color too."

"So you're not sure what triggers it?" asked Dr. Wood.

"It just happens. It's normal," said Victoria. "I hardly think about it."

"Is it just sight and sound?" I asked.

"Well, smell does it, too," she said, "but not as often as sounds do. Some smells have definite colors, you know," she gestured. "Take strychnine, for example. You know how it has that wonderful, rich pink smell?" she asked, assuming that we did. "The funny thing about strychnine is that it has exactly the same pink smell as my angel food cake. Now isn't that odd?"

"Odd" was only one of the thoughts that came to mind. Victoria's synesthesia was pretty complex, involving sound, sight, smell, and pain. Because her sound-sight combinations seemed the most strongly developed, we did some impromptu experiments in which she sketched the shapes of foreign words that I spoke out loud. We used German and Czech words, neither of which Victoria understood. Dr. Wood and I soon satisfied ourselves that the diagnosis of synesthesia was accurate. Victoria had "colored hearing" one of the most common types of synesthesia.

"You should do a little experiment and submit a paper to the INS," Dr. Wood prodded me later that afternoon.

"What's the INS?"

"The International Neuropsychological Society. Now you've got two interesting subjects, a man who tastes shapes and a woman who hears and smells colors. The INS has a meeting coming up in February. You should think about submitting a paper."

"I should?" I asked, unconvinced of the need to go to so much trouble.

"I'd bet they've never had a submission on synesthesia before."

"You think so?" I asked.

"I'm sure you could get a paper accepted," he assured me.

The Chief's suggestion to think up a "little experiment" was really a direct order to get with it.[10] I had three months before the INS submission deadline to whip up a proposal. Off I went to the medical library to see what others had written about synesthesia. A computerized bibliography search turned up zilch. This is not a happy result when you have to do a project on a subject about which you know next to nothing.

Since the computer database only went back to 1966, I tried the old-fashioned approach of looking it up in a book. I started with some bound volumes of the *Index Medicus*. This bibliography was begun in the eighteenth century by the Army Surgeon General's Office and was continued by the National Library of Medicine. I found some pretty old references. Fortunately, the eight floors of the medical library were extremely well stocked and contained many of the journals in question.

The references I had found in the *Index Medicus* sent me down into the third sub-basement. In the stacks, covered with dust and obviously unopened for decades, I found articles from the nineteenth and early twentieth centuries on synesthesia in French, German, Italian, and English. I eagerly skimmed through them in the dim fluorescent light. Just like the odd visual conditions I had found in old ophthalmology texts, every reference to synesthesia was in some authoritative but forgotten journal, often written by someone whose name was well known in medicine or psychology. These old papers described cases of synesthesia quite well, but none of them gave a clue as to what made it happen. Like Luria, it was description without explanation.

With what information there was decades out of date, I resigned myself to starting from scratch.

Chapter 8

▲ ■ ●

DOWN IN THE BASEMENT:
THE HISTORY OF SYNESTHESIA

Back down in the library sub-basement, I pried open the yellowed pages that revealed synesthesia's past. What a scientist ultimately wants is a theory that explains things, and what I was looking for was a theory to explain the brain mechanism of synesthesia. I was searching for that higher plane.

What I found instead was confusion. The archaic nineteenth-century language was not that impenetrable, nor were the naïve errors and anatomic speculations of what I found. Amusing phrases such as "a tangle of the optic and auditory nerve fibers" or the more flowery "echoing of the hearing nerve to the chromatic fibers" could be forgiven, considering how much our conceptualization of nervous tissue had changed in a century. Rather, the confusion stemmed from lack of agreement in what people meant by the word *synesthesia*.

Some authors clearly were talking about imagination, in which, for example, volunteer students *imagined* that words like "fiery" and "red" somehow went together appropriately with the music of a marching band. Other writers spoke of metaphoric speech, speculating on the origins of such commonplace sentiments that "yellow" was a "bright" color. Others accurately described synesthesia but apparently could not locate any real synesthetes; they erred in using non-synesthetic volunteers in their sense-to-sense matching experiments.

"What a mess," I muttered to myself. A good portion of what I read was nothing like the experiences of Michael or Victoria, or even of Luria's S. Those three did not stop to think about their sensations. They just happened.

I found a few papers that focused on *physical sensations*. In these cases, a stimulus in one sense would make subjects involuntarily experience something in another sense. They claimed they had "nothing to do with it," but that it just "happened by itself." Moreover, they were surprised to discover that anyone would consider it unusual, assuming that everyone felt this way.

No wonder history had been unable to explain synesthesia, I thought. People had used the same name for different experiences. It was like having a single word that stood for Greyhound buses, shoelaces, and treacle tarts. How could anyone be sure what you were referring to? No, this only confuses the issue, I thought. My first step would be to weed out cases that failed to explicitly refer to synesthesia as an *involuntary experience* in which the stimulation of one sense caused a perception in another. Later, I would have to pen a strict definition so that others would at least be certain what I was talking about.

No matter what subject science scrutinizes, the historical, descriptive, and experimental approaches are its three mainstays. Which method you use depends on your problem and the tools at hand. I would eventually apply all three methods to synesthesia, diving into its rich history first. No matter how a scientist approaches a problem, however, each method demands evidence, clear reasoning, testable hypotheses, a search for explanatory and predictive theories, and an effort to identify and avoid bias. These are the foundations of the scientific enterprise.

The first medical reference to synesthesia was about 1710, when an English ophthalmologist, Thomas Woolhouse, described the case of a blind man who perceived sound-induced colored visions. Even earlier, in 1690, the philosopher John Locke[11] had written of "a studious blind man who . . . bragged one day that he now understood what *scarlet* signified. . . . It was like the sound of a trumpet."

I was puzzled that neurologists, specifically, had hardly pondered synesthesia. It would have seemed a natural subject. It was mentioned sporadically throughout the eighteenth century, but in the nineteenth it attracted serious attention, scientific and otherwise. It particularly interested psychologists, artists, and natural philosophers. For example, in 1704 Sir Isaac Newton[12] struggled to devise mathematical formulas to equate the vibrational frequency of sound waves with a corresponding wavelength of light. He failed to find his hoped-for translation algorithm, but the idea of a correspondence took root, and the first practical application of it appears to be the *clavecin oculaire*, an instrument that played sound and light simultaneously. It was invented in 1725.[13,14] Charles Darwin's grandfather, Erasmus, achieved the same effect with a harpsichord and lanterns in 1790,[15] although many others were built in the intervening years on the same principle, whereby a keyboard controlled mechanical shutters from behind which colored lights shone. By 1810 even Goethe was expounding correspondences between color and other senses in his book, *Theory of Color*.[16]

I had soon found many cases of synesthesia recorded in both the scientific and general literature, as well as two books devoted to it. *Colored Hearing*[17] was published in French in 1890, while a German text appeared in 1927 called *Colored Hearing and the Synesthetic Factor of Experience*.[18] Most accounts emphasized colored hearing, which I discovered was the most common form of synesthesia.

This disproportion in the types of synesthesia was itself intriguing. The five senses of sight, sound, taste, touch, and smell could have ten possible synesthetic pairings: sight with sound, sight with taste, sight with touch, and so on. Synesthetic relationships usually operate in only one direction, however, meaning that for a particular synesthete sight may induce touch, but touch will not induce visual perceptions. This one-way street, therefore, allowed twenty permutations of sensory pairings. Given this number of possible pairings, I noted with interest that some senses, like sight and sound, were involved much more often than smell, for example. To persons endowed with colored hearing, sounds—especially speech and music—were not only heard but produced a visual mélange of colored shapes, movement, patterns, and brightness.

Some pairings had never been witnessed. While in one of Victoria's synesthetic combinations sight produced smell, I could find no cases in which smell itself was the trigger. And I only found one other in which, like Michael, taste induced synesthesia. In this instance, it was a case of colored taste.

Aside from Michael's own geometric taste, perhaps the strangest example of synesthesia I uncovered was "audiomotor," in which a fourteen-year-old boy positioned his body in different postures according to the sounds of different words.[19] Both English and nonsense sounds had certain physical movements, the boy claimed, which he could demonstrate by striking various poses. By way of convincing himself that this sound-to-movement association was real, the physician who described it planned to retest the boy later on without warning. When the doctor read the same word list aloud ten years later, the boy assumed, without hesitation, the identical postures of a decade earlier.

The addition of physical movement as a potential synesthetic response really meant that there were six possible components rather than five. Correspondingly, this implied thirty permutations. And yet only a handful of synesthetic combinations accounted for most of the cases. I made a mental note to remember the curious fact that some senses were privileged to occur in synesthesia, while others rarely, if ever, did. Perhaps anatomy would turn out to have something to do with it.

IDEAS OF SENSORY FUSION

By the nineteenth century synesthesia had gained the attention of a larger art movement that sought a sensory fusion. A union of the senses appeared more and more frequently as an idea in literature, music, and art. Multimodal concerts of music and light (*son et lumière*), sometimes including odor, were popular and often featured the color organs I had read about, keyboards that controlled lights as well as musical notes. Mrs. Astor, at her Beechwood mansion in Newport, took hold of this idea of sensory fusion by pouring expensive French perfume into the crystal cups of her candlelight chandeliers, so that her well-fed, besotted guests could experience yet one more sybaritic pleasure.

This was amusing, but went too far, I concluded. Examples such as this made me more aware that precise terminology was essential. I would use "synesthesia" only to refer to involuntary experiences. Calling deliberate contrivances such as Mrs. Astor's "synesthesia" only further confused the meaning of the word. Surely there was a difference in both the intent and experience of composers who wrote colored music as well, I thought. Arthur Bliss, for example, decided to write his 1922 *Color Symphony* based on the idea of synesthesia. He never claimed to have it himself; the project was just an intellectual exercise, his color choices arbitrary.

On the other hand, the Russian composer Alexander Scriabin

(1872–1915) specifically sought to express his own synesthesiæ in his 1910 symphony *Prometheus, the Poem of Fire*. It was written for a large orchestra with piano, organ, and choir. It also included a mute keyboard, a *clavier à lumières*, which controlled the play of colored light in the form of beams, clouds, and other shapes, flooding the concert hall and culminating in a white light so strong as to be "painful to the eyes." Its part in the score is written in conventional notation, with Scriabin providing his synesthetic code between notes, color, and shape. Keys also had specific colors. From the score's symbolic cover, to which Scriabin attached enormous importance, one gets a feeling for his obsession to express an almost spiritual mysticism through sensory fusion. He actually invented the "mystic chord," a series of five fourths (C, F$^\#$, B$^\flat$, E′, A′, D″) that is instantly familiar and that forms the harmonic basis in most of his works.

The first performance of *Prometheus* with lighting effects was March 20, 1915, in New York, just five weeks before the composer's death. Technical difficulties were insurmountable. As a compromise, which Scriabin rejected, the colors were merely projected on a screen above the orchestra. Technical reviews are given in *Scientific American* issues of the period.[20] A much more ambitious and all-embracing *Mysterium* was never completed. The work was planned to open with a "liturgical enactment," in which music, poetry, dance, colored light, and fragrance were to unite, inducing the worshipers to a "supreme, final ecstasy."

Vasily Kandinsky (1866–1944) was a synesthetic artist who perhaps had the deepest understanding of sensory fusion.[21] He had stopped objective representation in his paintings after 1911, being more interested in expressing a vision than in illustrating surface reality. Kandinsky was among the first to step off the well-beaten path of representation that Western art had followed for five hundred years, and his model to express his transcendent vision was music. He explored harmonious relationships between sound and color and used musical terms to describe his paintings, called them "compositions" and "improvisations."

Kandinsky studied piano for a time in Moscow, where he was captivated by the promise of artistic fusion he found in the Symbolist movement. His one-act opera of 1912, "The Yellow Sound" (*Der Gelbe Klang*), specified a compound mixture of color, light, dance and sound. The actual music was composed by Kandinsky's friend Thomas De Hartmann, who along with Kandinsky, Paul Klee, and

Arnold Schönberg, was connected with the avant garde Blue Rider group in Vienna.

Although what we think of as typical Kandinsky paintings were not produced until mid-life, they embodied ideas he held from an early age. With quantum theory and relativity published in 1900 and 1905, modern science was painting its own picture of the world, one that was as different from its classical predecessor as were Kandinsky's paintings from those featuring regimented one-point perspective from the Renaissance. Kandinsky absorbed the teachings of Theosophy[22] and Eastern thought, and the ideas he encountered in both scientific and esoteric writings confirmed a spiritual view of the world that he had held since his student days. In essence, Kandinsky's conviction was that art, if it was to portray reality, should not concentrate on rendering things but on an intuitive process that he exercised in abstract painting, and in which he believed the spiritual could be found. In 1910, he wrote *On the Spiritual in Art*, adjuring:

> lend your ears to music, open your eyes to painting, and . . . stop thinking! Just ask yourself whether the work has enabled you to "walk about" into a hitherto unknown world. If the answer is yes, what more do you want?

Kandinsky wished to push aside analytic explanations and move himself and his audience closer to the quality of direct experience that synesthesia represented. He grasped the axiom that creativity is an experience rather than an abstract idea in urging others to "stop thinking!" He knew that experiencing something is impeded by a mind that constantly analyzes what is there. I take Kandinsky to mean that the artist is off track if he asks, "What am I trying to communicate here?" just as the viewer whose first response is "What is this junk supposed to be?" also misses the point. Kandinsky's referral to, shall we say, the *purpose* of painting is similar to Joseph Campbell's later dictum that "what people seek is not the meaning of life but the experience of being alive."

A CLASH OF SUBJECTIVE AND OBJECTIVE VIEWS

The French have rarely shunned anything mystical, and I was not surprised to find them particularly fascinated by synesthesia.

Nineteenth-century French anticlerics were most eager to discredit theological positions about the soul and to translate spiritual mysteries entirely into psychological terms. An experience such as synesthesia, explained in terms of a concrete psychology, suited their purposes. The dawn of the art movement kept some of this flavor, and "know thyself" was a popular buzzword. Freud was also in intellectual fashion, and talk of libido and unconscious was chic. In addition to the rise of both psychology and psychoanalysis, the late nineteenth and early twentieth centuries saw too the rise of surrealism, symbolist poetry, automatic writing, and other explorations of the newly-discovered unconscious psyche.

Paris was a magnet for these movements, which were looking for ways to externalize fundamental feelings and the emotional imprinting of past experiences without being visually or verbally literal. The idea that intrigued people was that synesthesia seemed to have a direct link to the unconscious. This idea was tailor-made for that intellectual and artistic climate. Arthur Rimbaud (1854–1891), himself probably synesthetic, was a prime figure in the Symbolist Movement that helped awaken public awareness of synesthesia in his time.[23] His poems contained direct reference to synesthetic perceptions, the best-known example being *The Vowels*, written in 1871.

VOYELLES

A Black, *Ɛ* White, *I* red, *U* Green, *O* Blue;
some day I'll crack your nascent origins

A Hairy corset of clacking black flies
Bombarding agony pits of stench-ridden darkness;

Ɛ Frankness in steamers and pavilions, lances
Of Lofty glaciers, white kings, shivering umbels;

I Purple red spittle, laughter of sweet lips
Turning to rage or into penitent raptures;

U Cycles, viridian seas divinely shuddering,
Peace in creature-sown pasture, and peace
In furrows alchemy sows in intellectual foreheads;

O The great last Trump full of strange stridencies,

Silences crossed by Lucifer's trail of angels,
O—, in Omega, the violet beam of His Gaze!

—translated by Edwin Honig
by permission of the translator & copyright holder

Intellectual fashions rose and fell just as hemlines did. When exploration of individual mystic, introspective, and subjective experiences
eventually fell out of fashion, synesthesia also ceased to interest.
With time, attention turned away from qualitative experience to
more objective behavior—meaning that which could be quantified or
measured by machines.

Synesthesia had always been accepted as a real phenomenon by
the medical and psychological professions if only because it had been
independently and repeatedly noted by many investigators over a
two-hundred-year period. Historical developments and a shift in the
professional climate simply signaled its demise as a topic for study.
Experimental psychology began *assigning numbers to behavior* and
compiling statistics on groups of "subjects."

The individual person was no longer worthy of study. The individual, in fact, came to be seen as increasingly irrelevant, and so were
any feelings, opinions, beliefs, or personal background. These things
constituted the messy subjective part of the human psyche. The
so-called learned professions ran away from the content of individual
minds and focused for many years on external behavior, exemplified
by psychological behaviorism à la B. F. Skinner that reduced life to
stimulus and response. This chapter of human inquiry, which many
thoughtful persons had participated in, was like an intellectual dark
age and almost just as unbelievable when looked at in retrospect.
One contemporary even dismissed the existence of minds as
"invented for the sole purpose of providing spurious explanations."[24]

During synesthesia's heyday, mechanistic explanations—the search
for universal correspondences among the senses and an alchemical
zeal to find a formula for translating one sense into another—failed,
just as Newton's earlier attempt had, and by the 1930s synesthesia
was left with a reputation as a psychological quirk that no one could
explain in terms of current brain physiology. The search for a mechanistic solution during the nineteenth century was consistent with
the then common view of a clockwork universe based on Newton's
uniform laws of motion. It would still be two generations before peo-

ple would even begin to think in terms of Einstein's relativistic universe.

The Lack of Universal Agreement

One of the most glaring problems in trying to fathom a mechanism for synesthesia was the lack of obvious agreement about the parallel sensations that synesthetes perceived. That is, two individuals with colored hearing were not likely to agree on the color of a given sound. "Researchers" from earlier centuries had done little more than make lists of stimuli and synesthetic responses, followed by dismay that a pattern of correspondence was not obvious. The composer Joachim Raff, for example, agreed with the blind man quoted by Locke that a trumpet's sound was scarlet, but five other synesthetes with colored hearing insisted it was yellow-red, pure yellow, or blue-green. No matter what senses were involved, comparing two or more synesthetes gave the same results: there was no *obvious* similarity in their sensations.

Expecting synesthetes to agree on what they sensed was evidently a major mistake. Prior approaches that tried this had come up empty-handed. I concluded that there was not, in fact, any universal translation algorithm, and I accepted the empirical observation that the responses of synesthetes were individually unique, or idiosyncratic.

This issue continued to haunt me, however, because it seemed to have bothered earlier scientists so much. If earlier folks were bothered by a lack of universal agreement among synesthetes, then modern persons would likely be bothered, too. All synesthetes could tell us about were their perceptions. Maybe we were unable to see any similarity because we were looking at the wrong thing. I pondered the likelihood that the mistake of earlier times lay in comparing *perceptions* themselves instead of some earlier neural process that built up to a conscious perception.

By way of analogy, I thought of how apes and humans are alike although they hardly look like it. Much of their anatomy is alike, their brains are very similar, and, of course, there is little difference in their DNA.[25] But you do not have to go back all the way to DNA to see this similarity. If you are looking for relationships on the tree of life, or any family tree for that matter, members closer to the trunk resemble each other much more than members out on distant branches do. This is why family resemblances are more apparent in

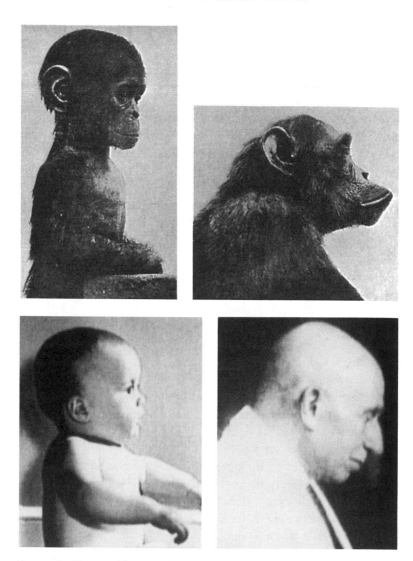

FIGURE 3. *Chimp and human infants compared to their adult forms. When comparing relationships in a family tree, similarities will be more apparent in members close to the trunk than they will be in members out on distant branches. (Chimp Photo from Næf A. 1926 Über die Urformen der Anthropomorphen und die Stammengeschichte des Menschenschädels. Naturwissenschaft 14:445-452.)*

offspring when they are young children than when they are grown-up. Even in the case of different species, a human infant and a chimp infant look remarkably alike, while adult members look wildly different (Figure 3).

All the intervening transformations between the eye and the visual cortex, for example, were possible candidates for neural processes that were closer to the trunk of perception than a completed visual image was. Modern physiology made it possible to trace with superb precision a visual signal from the moment light fell on the retina through its two dozen cerebral pathways. From this we could postulate detailed mechanisms of how derived aspects of an image, such as shape, color, location in space, direction of movement, or contour, are built up at the cellular level until we have a conscious visual experience. The intuition that earlier stages of processing rather than terminal ones were worth considering felt right to me.

Perhaps earlier scientists expected to find consistency in synesthetes' perceptions because of the consistency that all of us experience in everyday life. For example, we all agree that roses are red and violets are blue, that a square looks like a square, and that a banana tastes like a banana each time we eat one. We can recognize a piano by its sound and not mistake it for a trumpet or a baby's cry. We experience these things as consistent. But sometimes this consistency is an illusion. I thought of two illusions, called color constancy and colored shadows, so common in everyday experience that we take them for granted.

Color constancy is the illusion in which different stimuli look the same. The problem is that daylight is never the same. Its predominant wavelength, and therefore its color, varies as the sun travels through the sky. Scattering, reflection, and refraction by moisture and dust also change its color from moment to moment. Despite this constant change, a piece of white paper always looks white, an apple red and a banana yellow. People's complexions and clothing also look constant. The constant appearance of objects under widely varying conditions of illumination, intensity, and wavelength distribution is a well-known psychophysical issue and a central theme in understanding how we see.[26]

Because the predominant color of daylight changes from blue to red from sunrise to sunset, the same things viewed in the morning should look bluer than they do in the evening, when they should appear redder. People never notice this dramatic physical change, however. Instead, we perceive the color of an object as constant, despite continuous variation in both the intensity and color of the light. Even when the change is huge, as it is when we step from outside to indoors, colors still look constant. Our nervous systems attribute a constant color to an object, one different from what it "really is," and different from what the physical properties of the light should lead one to predict and perceive.

How ironic, I thought, that only artists, who are so often accused of looking at the world in a weird way, were adept at seeing these true dynamic changes in our visual landscape. They understood that color constancy is an illusion and often looked past it. Claude Monet, for example, was interested in decomposing light in much the same way a scientist might. How was it refracted, how was it reflected, and how did it impinge on the eye? One of his exercises was to paint the Cathedral at Notre Dame at different hours of the day to show how the color of the facade and the entire ambience changed. Even today, most people respond as Monet's public did, thinking that he was depicting his emotional response to the cathedral at each hour and not understanding that he was simply trying to paint it the way it looked.[27]

If color constancy is the illusion of different stimuli looking the same, then colored shadows is its opposite, the illusion of the same things looking different. All shadows have color, which is complementary to the illuminated side. For example, if an object is lit from the left by a colored light and from the right by a white light, then the shadow cast by the object blocking the colored beam will not be colorless, even though it "really" contains only white light from the right-hand source. It will have a striking color that is complementary to that of the colored light. If you photographed this setup using colored lights from different parts of the spectrum and then compared the photographs, the shadows would appear to be different colors, as expected. However, masking the colored light that surrounded the shadows would make them look identical.

To summarize these common illusions, the same stimulus can look different (colored shadows), or different stimuli can look the same (color constancy).[28] The flux of energy reaching our retinas changes constantly. The same is true for the flux reaching our other sense organs. Since our sense organs are energy transducers, our perception of what things "really" are should change accordingly. Instead, we are confounded by an illusion of constancy where none "really" exists.

My reverie had succeeded in turning my original assumptions upside down. The real question was not "Why don't synesthetes agree?" but "Why do the rest of us appear to agree so well when there is no compelling basis for agreement?" There were other philo-

sophical and biological arguments that flashed through my mind and
that seemed to support this conclusion. I decided to go over them
with Dr. Wood.

Having taken my historical foraging as far as I could, I returned to
the practical problem. Dr. Wood expected me to get a paper ready to
submit to the INS, and for that I had a pressing need for some fresh
data and an unambiguous definition of synesthesia. But first, I had to
think up a pilot experiment—a sort of fishing expedition—to see
how best to approach Michael and Victoria.

Chapter 9

APRIL 10, 1980:
"TASTE THIS!"

"It's an organic sphere."

I wrote his description down verbatim next to No. 6, which was Angostura bitters.

"With tendrils."

I appended Michael's weird elaboration in my notebook. "Organic?" I prompted.

"The shape feels like a living thing, see, which is why I say 'organic.' It's round but irregular, like a ball of dough." He cleared his throat. "The quinine, a few flavors back, felt like polished wood because it was so smooth. I guessed that one right away," he smiled. "But this sample is bitter in a different way. It's hard to describe."

He stuck out his tongue. "Gimme another squirt."

I picked up the syringe labeled "6" and squirted a dose onto Michael's tongue.

He shivered and squeezed his eyes together, holding the posture for a few seconds. "Ugh! Where do you get these things?" he choked.

"Tell me what it feels like," I prodded firmly.

"Totally different feel from the quinine," he said, rubbing his fingers together. "This definitely has an organic shape. It has the springy consistency of a mushroom, almost round," he said as he reached out, "but I feel bumps and can stick my fingers into little holes in the surface."

Michael closed his eyes and swept his hands through the empty air, feeling the shape of the bitter solution I had squirted in his mouth. Some of his associations sounded so funny that we had to stop for a good laugh. Still, I wrote down his every word exactly, no matter how ridiculous it sounded. Occasionally Michael amplified his comments with a sketch, but a picture was even less adequate than words in conveying what he actually felt. I thought of the three blind men feeling the elephant as I jotted down his description.

"There are leafy tendril-like things coming out of the holes," he said, pulling his hand through the empty air, "about six of them."

"This a mental image you see?" I asked.

"No, no," he stressed. "I don't see anything. I don't imagine anything. I *feel* it in my hands as if it were in front of me." I made notes.

"Squirt me some more," Michael said. "I'll be more specific." He stuck his tongue out and I gave him another taste.

He shivered past the bitter part and spoke quickly. "Yes, the round part comes first, with a spongy texture," he said, tracing a curve with both hands this time. "Then the shape develops—I feel the holes now," he said, closing his fingers. "Here are the strands. A little thread. It gets bigger, like a rope. If I pull my hand along one it feels like oily leaves on a short vine." He opened his eyes and sat up straight. "I guess the whole thing feels like a scraggly basket of hanging ivy." We laughed again.

"Aren't you going to tell me what you squirted in my mouth?"

"You know I can't," I answered. "It will confound the results if I tell you."

"The quinine and the sugar were obvious," Michael said, referring to samples we had already tried.

"It's impossible not to guess the simple flavors," I admitted, "but the less you know, the better. Experimental subjects are supposed to be 'blind,' officially speaking. Knowing could taint your responses."

What Michael and I were doing was called a pilot experiment, sort of a first try with no expectations in order to answer some elementary questions. What sorts of flavors did Michael respond to? Did smells also work, and if so did they cause different shapes than flavors? Should we use store-bought foods or make up chemical solutions in the lab? These were basic questions that had to be addressed before I could plan a more interesting experiment.

During our initial chats he explained how he felt and sometimes saw geometric shapes whenever he tasted food or smelled it. He felt some shapes, like points, throughout his whole body. Others, like the spheres of sweet savories, he felt only in his hands. Many shapes were in between, felt in his face, hands, and shoulders. What intrigued me the most was Michael's sense of grasping the shape, fingering its texture, or sensing its weight and temperature.

It was not surprising that he liked to cook and that he cooked by feel. He never followed a recipe but liked to create a dish with an "interesting shape." Sugar made things taste "rounder," while citrus added "points" to the food. He adjusted other seasonings to "make the lines steeper," to "sharpen up the corners," or to "make the surface stretch further back."

"With an intense flavor," he explained, "the feeling sweeps down my arm and into my hand. I can feel it as if I'm actually grasping it. There is nothing to see, of course, but I have a sense of movement." His synesthesiæ were usually pleasurable and sensuous. Rarely, he felt a "slap" or a "burning" in his face, or a "pricking" in his fingertips, "like laying my hand on a bed of nails. Mostly this happens with really sour foods."

I looked back through my notes. "Tell me what you mean by the shape developing."

"The shape changes with each moment, just as flavor does," explained Michael. "If you use sweet and sour sauce, for example, you taste the sweetness first, and then it becomes tart a moment later. The shape changes the same way according to how the taste changes. French cooking is my favorite precisely because it makes the shapes change in fabulous ways," he confided.

"If you taste the complexity of a French dish, with its layer upon layer in different locations in your mouth, the first taste is always a sweet one. It comes in flat and then becomes three-dimensional at the back of your tongue. It has this terrific movement to it. Sour tastes, on the other hand, have only two dimensions, either points or flat surfaces. Mostly, I like to cook with a single fresh herb or spice," he continued. "That way I can taste one glorious shape. I don't like too much going on."

Michael agreed that most people are metaphoric when they describe taste. "The whole vocabulary of wine sounds silly, you know, because most people have to describe one thing in terms of another. But for me, wine really has a shape. Describing some wine as 'earthy' is not poetic to me because it can literally be like holding a clod of dirt in my hand. On the other hand, I don't know what people mean when they say that cheddar cheese is 'sharp,' " he said. "That doesn't make any sense to me."

"The sensations are so hard to describe," he apologized. "If I sound metaphoric, I don't mean to be. I have to grope so for the right words."

▲ ■ ●

I understood that Michael's shapes were not metaphor. They were tactile perceptions that he explained by analogy to familiar objects. I would have to discourage this habit, however, and get him to state exactly the sensations he actually felt. With what I knew so far, I could formulate a plan for the next step of our pilot experiment.

Although both taste and smell worked for Michael, I decided to use flavors in our tests for two reasons. First, liquid flavors gave greater variety, made it easy to control strength, and let me also mix tastes. Secondly, the taste sense does not wear itself out like smell does. After sniffing flowers a few times, for instance, you can no longer smell them. Taste does not fatigue with repetition.

I now had a rough idea of what sorts of flavors triggered Michael's synesthesia and also what kinds of shapes he felt. His sensations were elementary things, like hard and soft; a smooth, rough, or squashy texture; warm or cool surfaces. Since the shape component of his sensations seemed to be geometric, I made up a circular diagram of shapes that ranged from completely round to completely angular. It served as an answer sheet for the first pilot experiment (Figure 4).

I hoped to see if Michael matched taste-to-shape in an orderly way, or whether it would be random like a normal person's. Ten trials of ten flavors, given in random order, are too much for anyone to keep track of. If Michael were just making up his story, his mapping should be as scattered as non-synesthetic control subjects over the hundred trials. If there really were some sort of link between his sense of taste and his sense of touch, however, some pattern in his matchings should emerge. The nature of such a link was not important to me yet.

I first gave Michael ten flavors in liquid form: (1) salt, (2) sucrose, (3) anise, (4) citric acid, (5) Campari, (6) menthol, (7) angostura bitters, (8) vanilla, (9) quinine, and (10) Karo syrup. Because I gave the one hundred tastings in a prearranged, counterbalanced order, I would also be able to tell if one flavor influenced another. That is, would a sweet shape change if it followed a sour taste instead of a salty one, or would it always remain the same, no matter what came before or after it?

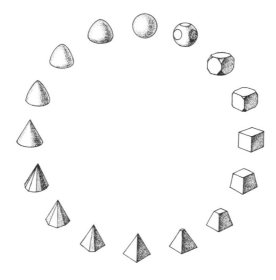

FIGURE 4. *The answer sheet for Michael's first pilot experiment. A circle of shapes varies continuously, with one shape changing into another. From "Synesthesia II: Psychophysical relationships in the synesthesia of geometrically shaped taste and colored hearing," by RE Cytowic & FB Wood, 1982, Brain and Cognition.* © 1982 *by Academic Press. Reprinted with permission.*

The first-pass results showed a definite pattern in Michael's mapping of taste to shape. Simple tastes, such as sweet or salty, generally made him pick far fewer choices than complex ones like anise or angostura bitters did. Certainly, he was not blindly guessing. It was clear, however, that the answer-sheet choices were inadequate to capture the lines, columns, and pointed shapes that Michael said he frequently felt. I revised the answer sheet into a figure-eight affair that represented much better the geometry he felt (Figure 5). With it he could quickly assign any flavor sample I came up with to one of the general shapes and then elaborate in words its texture, temperature, and other tactile qualities.

At the same time I was examining Michael, I was also going through a similar search with Victoria. In her case, I tried piano notes and other kinds of sounds on tape. The mapping of stimulus and response that we were doing was old hat in psychology. But it had a new twist with Michael and Victoria. Psychophysics, as it is

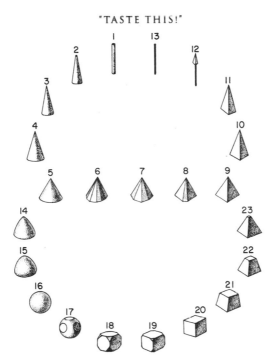

FIGURE 5. *The revised shapes for the taste-to-shape matching task now vary continuously in a figure-eight form. From "Synesthesia II: Psychophysical relationships in the synesthesia of geometrically shaped taste and colored hearing" by RE Cytowic & FB Wood, 1982, Brain and Cognition. © 1982 by Academic Press. Reprinted with permission.*

called, usually measures the threshold of a stimulus (how much it takes to perceive something), or its quantity (how loud, how big, how different from something else). The comparison of one sense to another, called cross-modal association, was not unheard of, although it was never meant to be taken *literally* as we were doing.

▲ ■ ●

"You know, if someone walked in on us they would think we were both crazy," Michael said during a break. We laughed. But we instinctively knew it was an accurate assessment of how others were likely to view our work. Synesthesia was just too off the wall for mainstream science in 1980. I had already seen my colleagues' reactions. If I had suggested using leeches I would probably have gotten a more receptive response. I found synesthesia fascinating. And I was also fascinated why nearly everyone else thought it was so taboo.

I told Michael about the reaction of my fellow residents.

"I'm not surprised," he said. "Imagine how I feel. All my life, people have thought I'm either crazy or on drugs. My parents thought I had an active imagination, and my friends thought I was being silly. That's why I was so embarrassed the night you came over to dinner. I hate it when people call me silly."

"Why?" I asked.

"Because I'm not being silly. I'm being me."

"When do you first remember having synesthesia?" I asked him.

"It goes back as far as I can remember, really. I can't recall a time when I didn't have it. I suppose it was in grade school, finally, that I realized I was different from the other kids. None of my friends knew what I was talking about when I mentioned the shapes."

"And your family wasn't supportive?"

"My childhood was not very happy," he said matter-of-factly. "I was left alone. I don't know when it began, but I always had an incredibly intense sense of smell. One of the happy things I do remember was visiting my grandmother in Arkansas. I would spend hours in her basement. It had the most wonderful combination of smells—the oil furnace, the root cellar, the storage room. I'd go for a visit and I would just be in heaven."

I asked Michael to sit up on the counter while I performed a neurological exam. "You said that it just happens. Is there nothing you can do to affect it?"

Michael thought a moment. "Nothing, absolutely," he answered. "The shapes happen to me. There is nothing for me to control. I suppose if I am tired or focused on something else, then I don't notice them as much. The morning, for example, doesn't stand out as a time when I seem to feel many interesting shapes. Perhaps at night, after a few drinks, I pay attention to them more."

"You may just be more relaxed," I suggested.

"Maybe."

I asked him to hop down so I could test his balance and coördination. "Our little pilot work is pretty boring," I said.

"And nasty," Michael interrupted. "Some of those tastes were horrible."

"You've been most coöperative," I continued. "I wonder what you expect to get out of this?"

"An answer, I suppose," Michael said with little hesitation. Michael agreed to my studying him because he wanted an explanation for being different. His father was an engineer, who explained how mechanical things worked. He and his father were the household repair team. Sadly, it was the only time they really spent together, the only way they communicated. "I guess I'd feel more comfortable if an authority could explain it. If Dad could explain how things worked then I hoped maybe science—that is, you—can explain how I work. I'd like to know that my synesthesia is for real." Michael still doubted his own experience.

His neurologic exam was normal. We sat talking over coffee at his kitchen table. Armed with a carton of bottles, vials, and syringes, I had come over to his house prepared to attack synesthesia with my head and with analysis. Over the next month I would become a frequent visitor, bursting in with bags and beakers, urging him to smell this or taste that. But I also came to learn a lot about Michael's inner life.

"You know, I studied botany in college," he said.

"Did you?"

"I love science. In fact, I wanted to become a physician," he beamed. We laughed at the irony of our respective situations. I was a chemist-turned-physician who had wanted to be an artist, and Michael was an artist who had wanted to be a physician. In my case, my father insisted that I follow in his footsteps. Michael's misfortune was getting bad guidance and believing that he lacked the aptitude for medical school.

"What a pity," he said. "I should have listened to myself." We nodded knowingly. "That's why I understand experimental method, you know, random order, hundreds of trials and all that," he said. "Science is actually pretty tedious, isn't it?"

"Yes," I agreed, "actual experiments are often tedious. Real science is based in the mind, not in machines. It isn't the big machines and computers they always show on television. The excitement lies in ideas and explanations, not in making measurements."

"Most people wouldn't agree," argued Michael. "People have this idea that theories are a dime a dozen. Look at newspaper editors, TV reporters, crackpots like Erik von Daniken. They spin amateur theories like spiders spin webs."

"All right then, silly ideas are a dime a dozen, if you want," I consented.

"The public can't distinguish what is plausible from what is nonsense or from the kind of claim that can only be accepted with proof," corrected Michael. "That's my point."

I agreed in part. "I think the popular view is that doing tedious, sometimes fruitless, labor is science. True, most scientists already work within or at the edges of established frameworks, so they don't spend their days theorizing. But they keep pushing the envelope and, ultimately, science is ideas; taking measurements is secondary. The second wrong idea is that science is big machines and computers. But ideas don't have to cost money. You can do science on a shoestring, like we're doing here."

"I hadn't thought of that," nodded Michael. "I guess you don't always need big buildings and enterprise. Look at Einstein," he laughed. "Relativity wasn't funded by government grants."

"And it changed our view of the universe," I added. "The third mistake people make is thinking that something is scientific if it can be proved. But the thing that actually makes a theory scientific is that it can be falsified."

"Are you sure?" asked Michael.

"Yes," I answered. "That idea was crystallized by Karl Popper, a philosopher of science." I thought about my plan for a moment. "What I think I want to do with you," I told Michael, "is propose the hypothesis that your senses work just like anyone else's. Then, I want to prove that hypothesis false."

I pondered the chasm, which seemed unbridgeable at that moment, between what science could "prove" objectively and what an individual knew subjectively. Given the inadequacy of traditional medicine to deal with patients who failed to fit its mold, I welcomed the opportunity for Michael to tell me that dark Karo syrup tasted "perfectly spherical, like hundreds of tiny, perfect spheres." I had seen no evidence to doubt his sanity. What little data I had from our fishing expedition suggested that his senses did not work like other people's. Michael was intelligent and talented. And he was courageous enough to tell me what he was feeling. I was pleased to earn his trust.

In time, attacking the problem of synesthesia with our heads eventually opened up both our hearts, and what that yielded was much more than either one of us expected. We would prove beyond a shadow of a doubt that synesthesia was real and that Michael was neither crazy nor silly. "It may be weird," he said years later, "but it's me. And that's OK."

▲ ■ ●

Chapter 10

DIAGNOSING SYNESTHESIA

Based on my reading and new first-hand experience, I set about to formulate clear-cut criteria for diagnosing synesthesia. The oldest criticism I had encountered against synesthesia being "real," or even a phenomenon worthy of scientific attention, was that it was subjective. That is, it had no external manifestations. It was a condition knowable only through the reports of those who claimed to experience it.

Many oddities in neuropsychology have been discovered by accident or through considerable coaxing of patients who feel embarrassed or ashamed by what culture has taught is "not real." Yet many people intuitively feel that research based on experiential reports is "unscientific" because it is not objective. They want a machine to reduce it to numbers.

Nonsense. Subjective experiences are the bread and butter of clinical neurology. They are, for example, the only means in classical neurology to assess any sensory quality, especially pain. There are abundant examples where verbal reports alone have changed our concept of brain function. The striking example is dreaming during the rapid eye movement (REM) stage of sleep. If no one had awakened sleepers during different EEG phases and asked them what was happening, the meaning of this dramatic EEG activity would have remained enigmatic, or it would have been assumed to be totally meaningless.

Admittedly, there were complex issues in trying to disentangle the how and why of synesthesia, but I believed it could in principle be understood just like any common experience, except that it had no shared reference. Experiential reports are often the neurologist's Rosetta stone. The stone has two sides, however. Experiential reports have unmistakably advanced neuroscience, so neuroscience should be able to help us better understand direct experience.

Aside from forgetting the role that subjective experience has played in the history of neuroscience, the criticism against subjective experiences is an empty one because many established medical conditions are entirely subjective. That is, they are diagnosed by symptoms alone. Headaches, dizzy spells, and temporal lobe epilepsy are obvious examples of ordinary conditions that have no external signs. The millions of patients who yearly visit their doctor to say, "I have these splitting headaches," are usually believed. No one questions immediately whether they are "really" in pain, yet nearly all will have normal exams and no objective signs. Despite the lack of objective evidence, doctors try different treatments without hesitation. (They order lots of tests, too, almost all of which are negative; this further lack of evidence does not dissuade them from treatment.) Headache, like all pain syndromes, is subjective. There is no way to prove that individuals do not feel the agony they say they do.

The meanings of words can clarify our thought processes. The word *diagnosis* literally means "through knowledge" (Greek *dia* = "through" + *gnosis* = "knowledge"). In diagnosis, knowledge aids the art of distinguishing one disease from another. (The word *disease* itself comes from the French *dès* = "from" + *aise* = "ease.") Syndromes make it easier to diagnose medical conditions by making patterns of symptoms and signs evident. The word *syndrome* is Greek and means "concurrence." Syndromes are oversimplifications that help to organize, weigh, and exclude information.

Rheumatoid arthritis, for example, is painful and associated with redness around the joints. Joint deformity can be seen by direct examination and on X-rays. A pattern of affliction exists because some joints are involved more often than others. The knuckles, for example, are affected before other finger joints are. Since the pathology of rheumatoid arthritis is inflammation, the symptoms and signs respond to anti-inflammation drugs. All these features constitute the syndrome of rheumatoid arthritis that distinguishes it from other kinds of arthritis. If deformed knuckle joints looked like rheumatoid arthritis but were *not* painful, for example, another diagnosis would have to be seriously entertained.

The associated features or characteristics of a syndrome are easily recognized by those trained to do so. Most often, these features come from the medical history, which is a story pulled out of the patient. Many might say, "I have these splitting headaches," but the story of a migraine is different from that of a sinus headache, which is different from that of a brain tumor headache, and so on. Doctors usually do not doubt that patients "really" experience what they claim because their stories sound familiar. Their features fit a syndrome.

In relation to synesthesia, temporal lobe epilepsy (TLE) is a more apt example than headaches or arthritis. Although these particular epileptics rarely have shaking convulsions as those with grand mal fits do, they do commonly have peculiar *subjective* experiences, such as a disordered sense of time; the experience of leaving one's body, called autoscopy; sudden and strong emotional feelings; and sensory illusions and distortions. From time to time they also experience synesthesia.

Temporal lobe epilepsy is common, affecting one in 9,600 persons. It is such a well-known condition that neurologists can *diagnose* it by the history alone. That is, the patient's story sounds familiar. Its pieces fall in place and fit a pattern. The diagnosis can be *confirmed* by prescribing seizure medication and making the symptoms go away. Finally, it can be *proved* by demonstrating characteristic waves on the electroencephalogram.

Although synesthesia is much rarer than temporal lobe epilepsy (about ten in a million), the stories of synesthetes sounded so similar that I concluded it could be diagnosed by features in the history. It could be confirmed by meeting five diagnostic criteria. And I hoped to prove this through experimental tests that would separate synesthetes from nonsynesthetes.

A given instance of a disease rarely shows all the known features of its syndrome. Disease X, for instance, may have ten textbook features in its syndrome. But a given case can satisfy the diagnosis with only three or four of them. Components of a syndrome are usually broken down into cardinal manifestations, without some of which the diagnosis is doubtful, and secondary manifestations, which are variably present.

Having looked at synesthesia's history, the first of the scientific enterprise's three mainstays, I now entered the second phase, that of description. I proposed five cardinal features for the diagnosis of synesthesia.

1. Synesthesia Is Involuntary But Must Be Elicited

Synesthesia is insuppressible. It happens to people and cannot be conjured up at will. The external stimulus that sets it off is often easily identified, yet not everything will cause a synesthetic reaction. Some synesthetes will respond to only a handful of stimuli, whereas others are sensitive to a much wider range of triggers. Synesthetes often say that their gift has existed as far back as they can remember and are surprised to discover that others do not perceive the world as they do.

In most cases, synesthesia does not interfere with everyday mental or physical activities, although it cannot be turned off or on, or willfully suppressed. If one is deeply engaged, then synesthesia may not seem so noticeable, whereas deliberately focusing attention on it in a relaxed setting may make it seem more vivid. Otherwise, the person cannot alter the synesthetic percept.

2. Synesthesia Is Projected

The parallel sense that is triggered is usually perceived outside the body rather than "in the mind's eyes." If visual, synesthesia is experienced close to the face. In other modes of sensation it is sensed in personal space—the space immediately surrounding the body—rather than at a distance.

3. Synesthetic Perceptions Are Durable, Discrete, and Generic

The associations of an individual synesthete endure for a lifetime. If a sound is blue, it will always be blue. If lemon is a pointed shape, it will always be so. This feature had been affirmed repeatedly, by others, by testing individuals without warning up to forty-six years apart with the same stimuli.

Saying that synesthetic perceptions are discrete means that given choices on a matching task, synesthetes pick only one, or a few at most, whereas non-synesthetic control subjects pick diffusely over the available selections. Discreteness also means that the sensations experienced by the synesthete have unique, "signature" qualities. We all recognize the distinctive sound of a piano because it sounds like a piano, not like a vacuum cleaner or a dentist's drill. As one subject explained, "The shapes are not distinct from hearing—they are part of what hearing *is*. The vibraphone, the musical instrument, makes a round shape. Each note is like a little gold ball falling. That's what the sound *is*. It couldn't possibly be anything else."

That synesthetic perceptions are generic means that they are never complex scenes. They are unembellished percepts: blobs, lines, spirals, and lattice shapes; smooth or rough textures; agreeable or disagreeable tastes, such as salty, sweet, or metallic. Synesthetes do not see cow pastures and temples, taste chicken soup like what mother used to make, or feel a sponge. It never gets that specific.

Symmetrical replication of the sensation is also common. For example, one line becomes four parallel ones, or a circle becomes concentric ones like ripples on a pond. Synesthetic percepts never go beyond this elementary, unembroidered level. If they did, they would no longer be synesthesia but rather well-formed hallucinations or figurative mental images of the kind we all have while daydreaming.

4. Synesthesia Is Memorable

The parallel sensations are easily and vividly remembered, often in preference to the stimulus that triggered them. "She had a green name—I forget, it was either Ethel or Vivian," says a woman named Diane. She confuses the actual names because they are both green, but she remembers the synesthetic greenness.

There is a strong link between synesthesia and photographic memory (technically called eidetic memory) or at least heightened memory (hypermnesis). Many synesthetes used their synesthesia as a mnemonic aid. The relationship between synesthesia and memory was best depicted in Luria's book, *The Mind of a Mnemonist*. His subject's memory, which was limitless and without distortion, was largely so because of the synesthesiæ that involuntarily accompanied every sensation.

5. Synesthesia Is Emotional and Noëtic

Synesthetes have an unshakable conviction that what they perceive is real. Their perceptions are accompanied by a "Eureka" sensation, the sense of the lightbulb turning on that comes with an insight (the "this is it" feeling). The invariable presence of such strong feelings of validity forced me to consider what contribution the limbic brain made to synesthesia.

The limbic brain, a structure much older than the cortex, deals with emotion and memory and provides the sense of conviction that individuals attach to their ideas and beliefs. The emotion and sense of certitude that accompany synesthetic experience made me think of that transitory change in self-awareness that is known as ecstasy. Ecstasy is any passion by which the thoughts are absorbed and in which the mind is for a time lost. In discussing mystical experiences in *The Varieties of Religious Experience*, William James spoke of ecstasy's four qualities of ineffability, passivity, noëtic quality, and transience. There are exactly the same qualities of synesthesia.

"Noëtic" is a rarely used word that comes from the Greek *nous*, meaning intellect or understanding. It gives us our world "knowledge," and means knowledge that is experienced directly, an illumination that is accompanied by a feeling of certitude. It breaks through surface reality and gives a glimpse of the transcendent. James spoke of a "noëtic sense of truth" and the sense of authority that these states impart.

> Although so similar to states of feeling, mystical states seem to those who experience them to be also states of knowledge. They are states of insight into depths of truth unplumbed by the discursive intellect. They are illuminations, revelations, full of significance and importance, all inarticulate though they remain; and as a rule they carry with them a curious sense of authority for after-time.[1]

My thoughts were now clear about what synesthesia was and was not. I could proceed to the next step, which was conceptual.

Most people think of science in terms of apparatuses, or machines, instead of thinking of it as a conceptual enterprise, a way of making your ideas clear to yourself. This is what leads to whatever apparatus, if any, you need to support or refute those ideas. This common view of science is very much like the common view of medicine, I thought: get a bunch of tests and see what they show.

Dr. MacArdle had adjured us that thoughtful diagnosis should lead to the judicious use of tests that could prove or disprove a diagnosis. For example, you do not order a CT scan "just because." Instead, you say, "I think a tumor is likely." When you get the CT scan, therefore, you ask, "Will the scan confirm my probable diagnosis of a tumor and, if so, will it distinguish between the possible types of tumor, A, B, and C?"

Similarly, as a scientist, you need to have a clear idea of what you want to ask before planning a thoughtful experiment to answer it. You can't just wildly try this and that and hope to see what comes up. Just as a doctor thinks about symptoms, the scientist ponders questions, mulling over *what* question to ask and *how* best to pose it. The first question I wanted to know about synesthesia was "Where is the link?" Where in the brain does the synesthetic joining of two senses take place?

Chapter 11

APRIL 25, 1980:
WHERE IS THE LINK?

"So what did you find out from the library?" Dr. Wood asked.

"There are no direct comparisons between synesthetes and non-synesthetes. I went back to the early nineteenth century and couldn't find a single instance."

"You mean no one ever compared how they perceive?" he asked, stubbing out his cigarette.

"That's right. I think we should be the first to do it."

"I agree," he said. "What else did you learn?"

"Most of it sounds pretty funny because the concepts about the nervous system are way out of date. Take the sensory reflex. That was my favorite," I grinned.

"What is that?"

"It's supposed to be a reflex, like the knee jerk is," I explained, "except it doesn't have any motor half. Instead of a motor response, the other half is hooked up to another sense. Pretty clever, huh?"

"The anatomy is imaginative," he laughed. "But you have to remember that psychological theory was terribly conflated back then," he said. I thought Dr. Wood should know since he taught a graduate course in the history of psychology. "Did the psychologists suggest anything interesting, or were they out in left field as much as the reflex people?" he asked.

I flipped through my index cards. Dr. Wood was not eating at the moment, so he paced around the perimeter of his office instead. "Here's the *association theory* from 1922," I called out. "It explains synesthesia as a chance association. If A suggests B, then A and B have been experienced simultaneously sometime in the past."

Dr. Wood rolled his eyes. "Anything useful?"

"I'm just trying to show how little there is to go on." I thumbed through the cards. "From 1895 we have the *emotional tone theory*, which emphasizes synesthesia's emotional force," I said. "It suggests that the stimulus and its synesthetic response have a common emotional background."

Given what we had learned from our first-hand cases, plus the dozens I had culled from the literature, we knew this could not be so. If it were true, the correlate would require that synesthesia be ubiquitous rather than rare, since emotion colors all sensation. Logic would require a pleasant color to trigger a mellifluous sound, an ambrosian taste, a warm feeling, the fragrance of a flower, and so on. We knew that synesthetic associations were specific, not sweeping as this idea proposed.

"Looking through it all, I sensed that shared meaning was considered the most plausible explanation in the past," I suggested. "You know, language is the link."

"How so?" Dr. Wood asked.

"In our everyday lives we routinely make imaginative connections among our senses. We say, 'I see what you're saying,' or 'This cheese tastes sharp.' "

"Those are metaphoric descriptions," said Dr. Wood.

"I know. We don't really mean that the cheese has a knife-like edge, but the best that earlier researchers could do was suggest that synesthetic experience might be an exaggeration of this habit of language. Accordingly, the synesthete is someone with an overactive imagination who takes metaphors literally. This theory, known as *semantic mediation*, simply makes synesthesia a special instance of the more general use of metaphors."

"What is the evidence for semantic mediation?" asked Dr. Wood.

"A single study. And even here they failed to use live synesthetes. They based their conclusions on earlier written cases and compared them with non-synesthetes who were asked to match, deliberately, one sensual quality to another."

"How do you mean?" asked Dr. Wood.

"They showed that both synesthetes and non-synesthetes match up low pitches with large, dark photisms."

"By photisms, you mean a patch of light?"

"Correct, a luminous line or blob. Anyway," I continued, "they judged that low pitches went with large, dark photisms; high pitches with bright, small photisms; and louder sounds with brighter and larger photisms. What they showed was that synesthetes tended to follow the same conventional trends that most people do when deliberately asked to associate different qualities."

"What is that supposed to prove?" Dr. Wood questioned, breaking his step around the room.

"That's my point. It doesn't prove anything. The quality of the two experiences is worlds apart. The synesthetes claim actually to see different sizes, brightnesses, and colors that are not in their mind's eye, whereas the normal people merely imagine that these 'go together' somehow. While these relationships are interesting, the problem is that they only apply to colored hearing, not any other kind of synesthesia."

"Yes, it isn't possible to generalize the results," agreed Dr. Wood. He hummed softly. "Suggesting that shared meaning is the link between one sense and another really doesn't clarify anything, does it?" He resumed his pacing, walking carefully. "I like your earlier suggestion," he announced, looking up. "We need to compare, directly, what is characteristic of how synesthetes perceive with what is characteristic of how non-synesthetes perceive."

"How do we do that?" I asked.

"First tell me more about the older theories."

I had been able to divide ideas from the nineteenth and early twentieth centuries into three categories, which I called sensory incontinence, linkage theories, and abstraction theories.

Sensory incontinence suggested that the nervous energy associated with one sense leaks to another part of the brain, much as a puddle of water sloshes about in a rocking boat. The idea was based on an analogy to motor activity in infants. Consider an infant reaching for a toy. It is easy to observe that the other three limbs, and even the trunk, involuntarily flex as it tries to grab the toy with its preferred hand. As the child grows up, the motor pathways mature and become insulated from one another. At this point, cross-talk between pathways stops, and the child can begin to grasp accurately while keeping everything else still.

Some involuntary movements remain even in adults, as when you try to touch your little finger to the center of your palm. It is impossible to keep the other three fingers extended straight while doing so.[2] The involuntary flexion of the other fingers is called synkinesis (Greek *syn* = "join" + *kinesis* = "movement"). Since synkinesis is seen in all infants, the reasoning by analogy was that synesthesia, meaning "joined senses," stems from an immature nervous system that is unable to prevent energy from indiscriminately leaking from one sense to another.

"In other words, it's a kind of atavism," interjected Dr. Wood, "a throwback, since there is a level of development in animals at which no distinction among senses is thought to take place."

"It suggests that nothing distinguishes one brain region from another," I added. "Sensory incontinence logically predicts severe birth defects and brain malformations. Synesthetes should be severely retarded if they survived."

"And synesthetic response to a stimulus would be indiscriminate as well," added Dr. Wood. "It wouldn't be specific." Another theory had to be discarded since its predictions were contrary to observed facts: synesthetes' responses were extraordinarily specific, and as a group synesthetes themselves were bright.

"Linkage theories is the name I have given to explanations based on the idea that there is something different about the hardwiring of synesthetes' brains. You know, the crossed wires idea. Actually it is more like a direct wiring that produces a one-to-one linkage," I said.

I walked over to the blackboard. "A hardwired link would produce a parallel color-to-pitch scale like this. Take aX, bX, cX . . .," I wrote on the blackboard. "If they were the vibratory frequencies in a musical scale, then a simple multiple like aXY, bXY, cXY . . . would predict the wavelength of the synesthetically induced color."

"Basically you mean a translation dictionary," said Dr. Wood.

"Precisely. This is what Newton and Erasmus Darwin tried to do when they searched for a physical translation algorithm between the senses. This one can't be true, either, because different synesthetes make different associations. There is no systematic one-to-one matching of tone and color," I said.

"Nor any other sense," added Dr. Wood. He started humming again softly and stared at the carpet. "What does that leave us with?" he wondered.

"The fact that a translation dictionary can't be made makes me doubt that language can really be the link, either," I said. "If language, which is the most abstract thing we know of, were the basis for synesthesia, then a bilingual dictionary to translate one sense into another should have been possible. There would have to be universal agreement."

"So it gives us more negative evidence that language cannot be the link," said Dr. Wood, "because synesthetes don't agree." He looked up. "That sounds like Kandinsky, doesn't it?" he asked. "Didn't his book *On the Spiritual in Art* talk about a universal dictionary for translating the essence of one sense into another?"

"I'm not sure," I answered. "He promoted the symbolist notion of transferring one sense into another, but I think we have to be careful not to confuse the idea of symbolism with concrete physical sensations, even though Kandinsky was synesthetic and elaborated the idea."

"The overlap is interesting, don't you think?" pressed Dr. Wood.

"Yes," I answered, "but I want to be clear that I'm talking about involuntary sensations. Artistic ideas about sensory fusion and deliberately contrived colored music are something entirely different." I fingered my index cards, noting how often art was mentioned in the history of synesthesia. "Still," I mused, "creative issues like that may be one of the larger consequences to which synesthesia points."

"How do you mean?" Dr. Wood asked.

"Synesthesia seems a little like an octopus, doesn't it?" I asked, "reaching its tentacles out into so many different areas. It's not just a bizarre neurological syndrome that affects a handful of people; I have an intimation that it is important to the rest of us—important in a way that I don't understand yet."

"You may have bitten off more than you can chew, Rick," Dr. Wood smiled. "So where does your poking around in the basement leave us? Are we any better off than when we started?" he asked.

I summarized our progress. "We have swept away a lot of material that goes against the observed facts. We have a clear definition of synesthesia. We've disproved old ideas, such as synesthesia being due to an immature brain, or possibly based in language. I'd say we had a clean plate to start over with except for one thing I don't understand."

"What is that?" Dr. Wood asked.

"Aristotle."

Dr. Wood lit up. "Aristotle? How did he get into this?"

"I'm not certain, but I came across the notion of an abstract idea possibly being the link, and it has something to do with Aristotle's common sense."

Dr. Wood chuckled. "You mean *common sensibles*," he corrected. "It's a philosophic term." I could see that I had piqued his interest. He stepped over to a bookshelf, cocked his head, and scanned the titles. "Aristotle is my favorite philosopher, Rick, did you know that?"

"Then I consider myself lucky. You can explain it to me," I told him. I looked through my notes. "The abstraction theories, which are the ones suggesting that language is the link, seem to fall back on Aristotle's common sensibles as the abstract ideas that carry meaning. Apparently the two things that synesthesia joins together have the same meaning," I said. "But I don't understand what a common sensible is."

Dr. Wood plopped a heavy green volume on top of the piles covering his desk. It perched precariously on a stack of term papers. "Here, in *De Anima*," he pointed. "Come look. This is where Aristotle says that you can distinguish between the particular and common senses." He read from the book:

" 'By an object peculiar to a particular sense, I mean one that cannot be perceived by any other sense, and in respect of which no deception is possible. Thus color is an object peculiar to sight, sound to hearing, and flavor to taste. Each sense judges the objects peculiar to it and is never deceived as to the existence of the color or sound that it perceives.' "

"However," Dr. Wood explained, "we can be aware of things around us in more than one way. We can see as well as feel the sizes and shapes of objects, see and hear the motion of bodies from one place to another, and can even tell whether that motion is slow or fast."

"So," I said, "an example of a common sensible would be movement, number, rest, size, or length. Things like that?"

"Exactly. Those qualities cannot be learned by a single sense only, because they are *common to several senses*. The common senses are not perceived through a special sense organ, but rather indirectly through the five particular senses."

Dr. Wood slowly drew his hand down the page, and again down the one after that. He wasn't humming now. "I'm still listening," I reminded him.

"I'm looking for a passage." He spoke slowly as he carefully turned the pages. "Aristotle believed that although the senses come from outside through separate channels of the five sense organs, they do not remain separate in our experience." He looked up at me to see if I was following.

"Our senses give us a world of objects of various sizes and shapes, in motion or at rest, and related to one another in space in a variety of ways. Our experience of these objects includes many more qualities—their colors, the sounds they make, their roughness or smoothness, and so forth. We passively receive these sensations through the five sense organs but are more active in assembling the seamless fabric of our experience. While sensory energy comes from the outside, sense experience of that outside world involves memory and imagination on our part."

I spoke slowly, trying to put it all together. "So, a unity results from the simultaneous perception of different qualities in one object?"

"Yes," exclaimed Dr. Wood, "Aristotle says 'as bile is at once bitter and yellow. This explains why we may mistake a thing, because it happens to be yellow, for bile.' "

I pondered this idea of simultaneous recognition.

"Ah! Here it is," exclaimed Dr. Wood. "Aristotle tells us about the nature of discrimination. 'Each sense discriminates the specific differences of the objects proper to it. Thus sight discriminates between white and black, taste between sweet and bitter, and so on. But we can also discriminate between white and sweet, and in fact between any two sensible qualities. How do we perceive generic differences?

" 'It is not possible to discriminate between white and sweet by a different sense for each of them; there must be one sense to which both of the compared qualities are discernibly present. Otherwise it would be like trying to establish a difference between two objects on the ground that you perceived one of them and I the other.

" 'Objects can be differentiated only where there is a single faculty to discriminate between them. In the case of white and sweet, as they are recognized as distinct, there must be a single faculty to affirm the distinction and hence a single faculty which thinks and perceives them both. We conclude from this that different things cannot be discriminated by a separate organ for each.' "

"I think I'm beginning to see the point," I said, "but I'd like to borrow your book and go over it a few times. Is Aristotle saying that there is an inherent similarity among certain things that is so fundamental to how we perceive that we take it for granted? Is that it?" I asked.

"In one sense you could say so," admitted Dr. Wood, "but look what happens when you pursue Aristotle's argument."

"The one about discrimination?" I asked.

"Yes. If you pursue Aristotle's argument, you become forced to say that the same faculty that *discriminates* white from sweet may also *fail to discriminate* them or perceive them as synonymous because of their shared qualities—hence producing synesthesia." Dr. Wood tapped his forefinger on the page. "I think we're on to something here," he said as the book slid off his desk.

He rescued it from the floor and then looked straight at me. "If you believe that an Aristotelian common sensible is the link, then you have to also believe that synesthesia occurs at the highest level of abstract processing in the brain."

"I see. That would be the hypothesis, yes," I concluded. "You could try to falsify that idea."

"I find it interesting that you use the world *level*," I said after a moment. "It has such a special meaning to neurologists." I was referring to the conceptual organization of the nervous system into vertical levels, each of which has its physical counterpart. The whole idea of the neurological exam is to test the working integrity of each level until you encounter loss of function, at which point you say that you have established the "level of the lesion." It is not possible to make a diagnosis without knowing this anatomic localization.

"Think about the functional levels from the bottom up," I suggested, "and ask yourself at which ones synesthesia could occur." I thought out loud, starting with the muscles, then the peripheral nerves themselves which innervate groups of muscles for movement and relay the sense impulses back to the spinal cord. These were unlikely candidates since they could not support the mental aspect of synesthetic perception. I was being systematic, working from the bottom up.

The spinal cord integrates movement and sensation. It has several levels itself, tested by banging the kneecap and elsewhere with a reflex hammer. Inside the skull this parcellation into levels continues at the brainstem, with its own subdivisions, and so on up until you reach the more complex levels of the cerebrum proper, which is concerned with higher mental functions.[3] This mental trip through the major levels of the nervous system emphasized to me that the majority of levels involve neural processes about which we can never be consciously aware.

"Obviously, synesthesia is a higher function," I said to Dr. Wood. "The question is just how high is it? At what level does it occur?"

Dr. Wood looked up from his book. "How does this relate to Aristotle?"

"The very act of organizing theories of synesthesia into categories, as we're doing here, sheds light on the possible levels at which it could operate, doesn't it? I mean, you can reason through a lot of them. For example, it would be helpful to know whether a visual synesthesia is produced in the layers of the retina or whether it's in the most developed part of the visual cortex."

"Or maybe somewhere in between," suggested Dr. Wood.

"Yeah, or somewhere in between. Our approach should be to ask where the level is. Once we know that, we can ask what the nature of the link is." We both thought about conceptual levels over the next few weeks. The abstract part of scientific thinking had begun.

Chapter 12

▲ ■ ●

PAINTING THE CEILING

There is a view of creativity called the "messenger of God" theory, which proposes that finished works are implanted, whole, in the creator's mind. The messenger of God theory points to people like Mozart, who claimed to hear completed symphonies in his head that he had only to write down.

Another view of creativity, consistent with the experience of numerous people, says that instead of being suddenly inspired geniuses, individuals who are called creative are really more like a dog with a bone. They refuse to let go of an idea. They stick with it for a long time. They mull over the problem at their workbench as well as in the most mundane places. They chew on it just as a dog chews on the same old bone for hours.

The bone image actually has a lot of life to it. A dog guards a bone safely between its paws when not actively chewing it; a lot happens to nurture a creative idea too, even when you appear not to be working on it. A dog will pick the most tasty bone, one with a lot of bulk and some meat still on it; creative people find no challenge in problems whose answers come easily, preferring to sink their teeth into something meatier.

The true marks of creativity are: (1) an ability to sense which problems are likely to yield results and so are worth tackling, (2) confidence that you can solve the problems you single out for solution, and (3) a dogged persistence that keeps you going when others would give up. Creativity does not result from mysterious visions that come in dreams, or from fortuitous circumstances. Creativity and persistence are synonyms. Constantly thinking about the problem, consciously and unconsciously, maximizes the possibility that any chance occurrence is likely to be useful in solving it.

▲ ■ ●

I was painting the ceiling of my tenant's apartment in the basement of my North Carolina home. My French horn player had got married and moved out. Fortunately, having Salem College, the medical school, the School of the Arts, Piedmont Bible College, and more meant no shortage of ready tenants. Renting basements, attics, and spare rooms to students was a cottage industry in Winston-Salem. Its help in paying my mortgage was welcome relief since I earned less than $11,000 a year as a resident physician.

I dipped my roller, firmly screwed to a ceiling pole, into the tray of Carolina Coatings "ceiling white," a good local product that was surprisingly cheap. I pressed it in against the stained pressboard ceiling tiles, making sure the beveled joints where the tiles came together got their share of paint as I rolled it along in time to Texaco's Radio Opera House, turned on loud upstairs. Live from the Metropolitan Opera, Cilèa's *Adriana Lecouvreur* wafted down to the basement and lightened the work on my Saturday off.

The paint was thick like mayonnaise and brilliant white like whipped cream. I should have asked the man why the ceiling kind only came in a single ground-zero shade when you had two hundred choices of white for the walls. Its brilliance reminded me of the stuff that I squeezed out of those Grumbacher tubes for my mother. I stopped to refill the roller when my eye caught a dried patch I had finished a while ago. Its bright vellum sheen had faded and its color had darkened.

"What happened to color constancy?" I asked myself. As Adriana hit a high note, the sun came out and heightened the difference between the wet and dry parts. Having brushed up on the particulars of color constancy, I was disturbed that my ceiling refused to stay constant. I wiped my hands on a wet rag. The fresh paint came off, but the dried stuff on my palms and around my fingernails stuck fast.

In a flash, I saw my faulty reasoning. Color constancy meant that identical things looked the same under changing conditions of illumination. But the wet and dry paint were not identical: the patch I finished an hour earlier was physically different from the spot I had just rolled on. It physically changed as the solvent evaporated and the pigment hardened. I was relieved that the laws of the universe had not been violated by my tenant improvement. As I looked at my ceiling, I knew that Aristotelian common sensibles were like the dried part. By analogy, the dried paint was like an abstraction or a distilled essence of the wet paint that used to be there.

My mind shifted and raced through a catalogue of examples that all demonstrated how a sense perception may be distilled to an abstract essence. I imagine I am looking at some sensuous, physical object sitting in the driveway, baking in the hot sun. Slowly, parts of it evaporate, carried away by the breeze, and it collapses into a two-dimensional pancake. Soon nothing is left behind but a residue that is some abstract essence of what the direct sensation once was. The scene quickly changes to my backyard, littered with objects. A plastic clothesline, a garden snake, and a bowling lane evaporate away, leaving length behind. A mantel clock and a jet plane passing overhead en route to Europe do the same: time and distance are also lengths. Now a picket fence and a bishop's miter dry out until only a point remains. A straight pin and the prick it makes on your finger is a point. I see a heated argument between two people, their words colliding until its climax makes a point. Concepts of transition between one state and another, the point where maximum reverses toward minimum, all these are embodied in the dimensionless existence of a Cartesian point. I imagine Michael Watson eating an orange. Somehow its taste has this pointedness, which he is able to feel. These were examples of abstract, common sensibles.

I carefully guided the roller around the ceiling fixture, squinting against the two bare hundred-watt bulbs. I watched the strips of paint crisscross over old, brown water stains. "Now you see it, now you don't," I mumbled. I lowered the roller and looked at where the stains had been. The water spots were like unconscious ideas, really there but hidden from view by the new paint. I turned to look again at how the dried section had changed color. Synesthesia isn't like this, I thought. It's not the residue that is left after something has evaporated away. Synesthesia was rich, sensuous, direct, and immediate. It was real to those who perceived it.

Upstairs, they were taking Adriana away to prison; she sang with dignity and defied the Parisian aristocrats. My pulse rose with excitement. Yes, synesthesia was almost the *exact opposite* of Aristotle, I suddenly realized. It was not a subtractive experience at all. It was not like pouring the experience of touch, taste, and smell through a coffee filter that held some qualities behind and only let abstract ideas like length, pointedness, or depth pass through. No, synesthesia was an additive experience. It combined two or more senses into a more complex experience without losing their own identities.

I thought of mixing up different colored marbles in a jar, but quickly discarded that example. There had to be an interaction, or blending, of the particles. The additive experience of synesthesia was more like cooking, yes. It was more like pasta primavera, in which you have green and white noodles mixed with different vegetables. You can still discern individual parts in the dish, but they blend together into a new and more complex experience.

A burst of shouting and applause snapped me out of my reverie. It was intermission time on the opera broadcast. It was a thrilling performance.

▲ ■ ●

My musings in the basement had clarified the question I wanted to ask: Is synesthesia something abstract and malleable, like language or Aristotle's common sensibles, or it is something direct and invariable, hardwired like the knee-jerk reflex?

Just as I had considered the analogy with synkinesis earlier, I now thought of another motor reflex. While it turned out to be only superficially similar, it fortuitously centered on the topic of levels that Dr. Wood and I had discussed. The process in question had to do with reflex facial movements that newborn babies make in response to simple tastes. Its technical name was the gustofacial reflex, from the Latin word for taste, *gustus*.

In the gustofacial reflex, sweet tastes evoke a smile, bitter tastes produce a look of disgust and sticking out of the tongue, while sour tastes make the newborn purse its lips. That is, different tastes produce fixed and stereotyped facial expressions. Unlike synesthesia, the gustofacial reflex is universal, producing *identical* responses in different infants.

The gustofacial reflex is an invariant behavior controlled by the lower brainstem. Ethologists such as Konrad Lorenz call it innate or instinctual. The pattern of such an inherited motor coordination does not change. It is resistant to exhaustion by repetition and appears in a wide assortment of mammals. The gustofacial reflex also shatters our assumptions about discrimination. Taste is well-developed and functions long before birth, the adult form of the taste bud clearly visible in human embryos by the fifth gestational month. The gustofacial reflex is routinely present in newborns, even in anencephalic monsters—children born without any brain, only a spinal cord and brainstem.

The striking point of the gustofacial reflex is the ability of the lower brainstem to *discriminate* between sensory signals and "decide" that some events are welcomed by the organism while others must be rejected as unpleasant or harmful. People are inclined to believe that discrimination between good and bad is a cognitive function based on life experience, learning, habit, and an emotional attitude. This is not so. The ability of brainstem structures to discriminate at an unconscious level made me see that it was possible for synesthesia to operate as far down as this lower level of the nervous system.

It seemed I had now cast out two anchors that would limit how far I could traverse the levels of the nervous system. Reasoning through the gustofacial reflex, I concluded that the mechanism of synesthesia and its link would have to be above the level of the brainstem. (It would presumably not be *at* this level given that synesthetic responses were idiosyncratic rather than identical as in the gustofacial reflex.) Similarly, evidence gathered so far about language indicated that synesthesia would probably not occur at this highest level of abstract brain processing. The link had to be somewhere in between.

▲ ■ ●

"It's like this cone," I said, drawing one on the blackboard. "The link is either high, low, or in between," I said, sketching the possible routes that synesthetic associations could take in the brain (Figure 6).

"It depends on the fact that synesthesia and Aristotelian common sensibles are opposite things," I told Dr. Wood. "That is what will pin down the level at which synesthesia occurs."

"How so?" he asked.

"What is the quality of synesthesia?" I asked rhetorically. "It's concrete, a direct experience that people feel, and taste, and touch. It has no *meaning*," I emphasized. "Remember how people claim that synesthesia helps them remember things? What they really recognize as familiar is the synesthetic sensation."

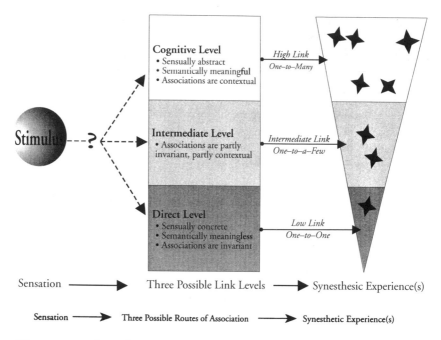

Sensation ⟶ Three Possible Link Levels ⟶ Synesthesic Experience(s)

Sensation ⟶ Three Possible Routes of Association ⟶ Synesthetic Experience(s)

"You mean that they recall the color, or whatever, more than the fact attached to it," said Dr. Wood.

"Right. Suppose you meet Ethel and see a green blob. The next time she comes around you don't say, 'Oh, it's Ethel,' you say, 'Hey, it's the green blob.' You remember the sensation better than her name. When you find out her name again, then you're sure because the synesthesia that goes with it is exactly the same as it was the last time."

Dr. Wood nodded. "An interesting idea. It is the *sensation* that is memorable, not the name. The name is just semantic baggage attached to it."

"Luria's patient S shows this very clearly," I said. "Despite his prodigious memory, he sometimes made mistakes. Those mistakes were not in reproducing what he was asked to remember incorrectly but in *omitting* some items from large series. He would use the method of loci, an old memory trick, by placing his synesthetic images throughout an imaginary town and then reading them off as he mentally strolled through it. When he did omit items, his explanation was not that he 'forgot' but that he didn't *see* them as he walked through his imaginary town."

"I remember reading that," said Dr. Wood, standing up. He went over to the bookcase. "His apparent errors of recollection were really errors of perception." He took down Luria's monograph and searched the pages. "I reread this since our last chat," he explained, looking at me over the top of his glasses. He flipped to a dog-eared page. "Here it is," he announced. "I remember this because it was so peculiar. His explanation for omitting items as he recalled them from a list was that he placed his synesthetic image in a spot where it was difficult for him to 'discern.' A white shape against a white wall or a dark object in a corner were omitted because 'he would simply walk on "without noticing" the particular item.' "

We found another relevant passage in which S said, "Sometimes if there is noise, or another person's voice intrudes, I see blurs that block off my images. It's these blurs which interfere with my recall."

"See, it's all sensation. There isn't any meaning," I said. I directed his attention to the blackboard. "Look here at this diagram. You have a stimulus on the left side that triggers synesthetic perception(s) on the right. The link can take one of three routes. If the link is at a low, direct level, then the stimulus-response combination should be invariant. It would be as hardwired as the knee-jerk reflex. You hit the knee cap and you get the same reflex over and over. It's a physical thing that doesn't have any cognitive meaning."

"I'm with you," he said.

"The context of the stimulus won't matter either. Assume I play a set of musical notes for Victoria and she says that 'A' is red. What I play before or after the 'A' won't alter its redness. A stimulus and response that does not vary with the context has a one-to-one link that should be at a low level in the nervous system," I said. "We should look at things like the retina, the ganglion cells in the ear, or the brainstem. Things that far down."

"All right," Dr. Wood said. "What about the high level, the one you've labeled 'cognitive'?"

"The high level is loaded with meaning. This is where Aristotle and cross-modal associations in normal people come in," I said, pointing to the widest part of the inverted cone.

"Explain cross-modal associations," Dr. Wood asked.

"Take a two-year-old child. You show the kid something and then put him in the dark with a bunch of objects. By touch alone he can pick out and recognize as identical the object he saw earlier. That's a cross-modal association, a human ability that even young children have."

"OK."

"We have known for a long time that the ability to make cross-modal associations is the foundation of language. Monkeys can't make them. In non-humans, the only readily established sense-to-sense associations are those between an emotional stimulus, like pleasure, and a non-emotional one, like vision, touch, or hearing. Only humans can make associations between two non-emotional stimuli; because of this we can assign names to objects.

"Let me go back to the monkey," I said. "It can establish an association between the sense of taste, say for a banana, and the emotional limbic brain, which registers 'yummy.' Anatomically, this is the only kind of connection possible in non-human brains. Because of the way that the human brain developed, humans can make a cross-modal association between a non-limbic sense, such as the taste of a banana, and assign it the abstract name of *banana*. From here, other abstract facts can be associated linguistically: bananas are high in potassium, they are picked by exploited migrant farm workers, they wound up in Carmen Miranda's hat, and so on.

"What I am saying is that cross-modal associations are the foundation of language. According to the standard view, language is the highest type of cross-modal association and especially depends on the tertiary association cortex and linkages between one part of the cortex and another. The entire process occurs in this youngest part of the brain."

"I follow you, but what is the relation to where synesthesia is associated?" Dr. Wood asked.

"Cross-modal associations are a normal part of our thinking, although they occur at an unconscious level. In synesthetes," I explained, "it is as if these associations poke through to awareness, like the sun poking through dark clouds so we can see it and feel its warmth. Even on the cloudiest of days, however, we know that the sun is in the sky even if we can't directly experience it.

"Even though we discern what we hear and see as distinct events," I continued, "experience also shows that we can integrate them in forming thoughts about what these sensations bring to our brains. This integration occurs at a level about which we are unaware. A small number of the human population, called synesthetes, act as if there is a conscious mixing of some of these sensory channels, as if a normal perceptual process that is usually hidden has somehow become bared to their consciousness."

I pointed to the cone on the blackboard. "Look at the wide part here," I said. "If the link for synesthetic associations occurs at the highest level of neural processing, then it will be abstract, like language or an Aristotelian common sensible. Associations at this high level are similar to metaphor that humans use innately. In this case, synesthetic perceptions should be rich in semantic meaning and will have lost all of their direct sensual properties. The experience will no longer be concrete, but abstract."

"Context will influence the experience as well," Dr. Wood added.

"Yes. There will be huge context effects. Let us assume in this case that Victoria perceives high notes as red. If I play her a bunch of musical notes in which 'A' is the highest, she will perceive 'A' as red. However, if I play her a second set of notes in which 'A' is the lowest of the group, then its color will change because its relative 'highness' has been drastically altered by the context."

"This is excellent strategy, Rick," nodded Dr. Wood. "What happens in the middle part of your diagram?"

"It is possible that the link is at neither the highest nor at the lowest level. In that case, it must be something in between. The cool part of this scheme is that the low, high, and in-between scenarios all predict consequences that can be verified or falsified by experiment. Given a particular stimulus, the number and scope of responses that a synesthete will make to it will be quite different according to the three possible routes of association."

"Go over them once more to make sure I have it straight," Dr. Wood said.

"Each of the three possibilities predicts whether the context of the stimulus matters," I explained. "Obviously we'll want to alter it systematically and see how Michael's and Victoria's responses change. If the link is at the lowest possible level, you'll always get the same association and there will be absolutely no contextual influence. But if the link happens at the highest level, then the stimulus should yield *more than one association*, all of which presumably share the same meaning that the stimulus has. The high-level link is a one-to-many link, producing synesthetic associations that should be no different from those of non-synesthetic control subjects. The third possibility of an intermediate position predicts a small amount of variation according to the context of the stimulus, but which still yields an association that has little, if any, semantic meaning."

Dr. Wood nodded. "Yes, good strategy."

The submission deadline for the INS was approaching, and I had my orders to devise a paper. "I guess I had better get busy with my experiment."

Chapter 13

▲ ■ ●

SUMMER 1980:
BRINGING THINGS TO A CLOSE

The vagaries of real life rarely allow for clean endings. I was midway through the experiments that constituted the third phase of my investigation into synesthesia when my fellowship ended and it was time to move. I had accepted a year-long position at the George Washington University Medical Center in Washington, D.C. Michael came up for a visit so that we could finish our work there.

As much as I had loved wilderness camping in Linville Gorge or rock climbing at Pilot Mountain, it was time for me to leave for a metropolis. I had fallen hard for Washington and had known for years that it was where I wanted to settle down. The excitement I felt during my visits there perhaps convinced me to return to urban living. Still, I had mixed feelings about leaving. An erstwhile carpet-bagger from New Jersey, I had made North Carolina my home for ten years. During that time, I picked up many colorful idioms but never the accent, as my friends reminded me. "You talk like us Rick, but you don't sound like us."

Even though Washington was below the Mason-Dixon line, it was nothing like the South I had come to know. Perhaps to retain a touch of country, I got an apartment on the edge of the National Zoo. You could hear the lions roar at sunrise, while at dusk came the haunting, blood-curdling cry of the peacocks as they swooped up into the branches for the evening. Best of all was a view that looked down over a mile of trees cutting a swath through the nation's capital.

Despite the sundry and dramatic changes, I had not lost interest in synesthesia. To prove or falsify the conceptual possibilities I had discussed with Dr. Wood, I devised a direct comparison of the characteristic ways in which synesthetes and non-synesthetes performed an identical task. I asked Victoria to match colors to twelve musical notes on tape; Michael's job was to match seven taste mixtures to twenty-three possible shapes. In addition to the standard controls of non-synesthetic persons of the same age and sex as Michael and Victoria, I employed two additional ones. These subjects were chosen for their professional response to taste, shape, color, and sound. Michael's extra controls were a chef and a carpenter; Victoria's were a visual artist and a professional portraiteur.

None of the control subjects had ever heard of synesthesia, and all approached their task with relish. Unlike my medical colleagues, these people were dying to know more about it. Until I had finished the experiments, however, I couldn't tell them anything. I instructed them that while they were engaged in the matching task, they should focus on the sensation of the stimulus, whether it was a musical note or a taste, and pick one of the color or shape choices from the answer sheet. They had to make a choice, I told them, and decide what went best with each stimulus. One of Michael's controls, a medical student named Tom, had to be replaced because he could not imagine how taste and shape could go together. An earnest but dull fellow, he couldn't even conceive of making matches. The others had no difficulty.

The actual task of the experiment was straightforward, although we had to go through each stimulus hundreds of times in order to answer my two pertinent questions: what is the level at which the synesthetic link operates, and is that link invariant or relative? By relative I meant, is it influenced by the context of the stimulus? I could answer these questions by systematically manipulating the stimulus, both by sweeping it through a *range* from low to high and also, within a given group of stimuli, by changing the *order* in which they occurred.

The pilot studies showed that flavors in solution for Michael and musical notes for Victoria were reliable and easily administered stimuli that consistently caused them to experience synesthesia. In her pilot experiment, Victoria initially chose her colors from an array of Munsell color chips, an industrial color standard. Earlier research by others, and by myself during the pilot studies, showed that color names were equivalent to physical samples, so I did away with the chips and used an answer sheet with color names printed on a single line.

Black Blue Brown Green Gray Orange Pink Purple Red White Yellow

Table 1 shows the arrangement of taste and sound stimuli, cleverly similar in that each contains a low part, a high part, and an extended range that incorporates both extremes. Since the concepts of "low" and "high" are more familiar when talking about musical notes than tastes, let me start with how the three tapes were prepared for Victoria.

Tape I tested the low part of Victoria's musical range. It consisted of twelve repetitions of the twelve notes between D-flat, next to middle C, and the C one octave above it. Each repetition contained the same notes in a different order. The order of the first set was random, but the order of each subsequent set was determined by a counterbalancing table to ensure that each note followed every other note once and occurred in each serial position once. Victoria therefore heard each of the twelve notes twelve times, for a total of 144 separate notes on each tape.

Tape III explored the high part of the musical range. It was identical to Tape I except for being one octave higher. Tape II tested the entire range by using the same pitches as in the low and high octaves, but spaced them out over four and a half octaves instead of one. This extended range also contained twelve repetitions of twelve notes counterbalanced as before.

Twelve lines of color adjectives, corresponding to one repetition of twelve notes, were printed on a single page. Thus twelve pages constituted the answer booklet for a single tape. Victoria and her three controls were instructed to mark the color that was perceived or imagined in response to the tone on the tape.

The shapes used for Michael's matching task were the ones in the figure-eight arrangement that systematically changed from a point, to a cylinder, a cone, a sphere, a cube, and back again to a point. I made thirteen solutions of equal concentration that ranged from a completely sweet taste (sucrose) for solution No. 1, to a completely sour taste (citric acid) for solution No. 13.

The solutions in between contained different proportions of sweet and sour. I tested Michael with three ranges of taste in the same way I tested Victoria with three ranges of pitch. The low range used

Table 1. "Low" to "High" Arrangement of Taste and Sound Stimuli											
Gustatory Synesthesia											
Sweet							Sour				
Experiment I											
1	2	3	4	5	6	7					
Experiment II											
					7	8	9	10	11	12	13
Experiment III (Extended range)											
1		3		5		7		9		11	13
Auditory Synesthesia											
Low							High				
Tape I (A-440) Low											
D♭ D E♭ E F F♯ G A♭ A B♭ B C											
Tape II Extended range (4 octaves)											
D♭ F A⁻²²⁰ D♭ F A⁻⁴⁴⁰ D♭ F A⁻⁸⁸⁰ D♭ F A⁻¹⁷⁶⁰											
Tape III (A-880) High											
						D♭ D E♭ E F F♯ G A♭ A B♭ B C				

solutions No. 1 (pure sucrose) through No. 7 (a 50:50 mixture of sucrose and sour acid). This low end was tested using seven repetitions of the seven tastes arranged, as in Victoria's case, by a counterbalancing table.

I probed the upper, or sour, part of this flavor spectrum by using solutions No. 7 (the 50:50 sucrose/sour acid mixture) through No. 13 (pure sour acid). Finally, the third run explored the whole range of tastes by using the odd-numbered solutions only. In each of the three runs, Michael and his control subjects tasted seven repetitions of seven tastes; these seven tastes sampled the low, high, and extended ranges in different runs.

I had cast the hypothesis that Michael and Victoria would be no different from the control subjects, a position that supposed that they were, in fact, making up their associations. If this hypothesis were true, then synesthesia would not be "real." Making it impossible for them to keep track of each stimulus and their response to it was part of the strategy to demonstrate this hypothesis false. Here was the difference between science as an idea versus science as observation. Victoria and her controls listened to four hundred and thirty-two notes while Michael and his gang tasted one hundred and forty-seven separate samples. This repetition was a tedious, but necessary, chore to see if my concept of how synesthesia worked was correct or not. Could I falsify the hypothesis that they were no different from control subjects?

Because I had put her stimuli on tape, Victoria worked alone. The fact that her experiment ran automatically did not lessen my tedium of making it or hers of executing it. Having to administer Michael's stimuli myself gave the two of us a chance to talk further.

"This wasn't what I expected at all," Michael admitted, after one of the trials. "This isn't exciting," he complained.

"Shall I throw you into a cyclotron and turn it up to 'tilt'?" I asked. "You want sparks and flashing lights, is that it?"

Michael laughed. He leaned back and wrinkled his brow. "Actually," he said, "I do want flashing lights. I want probing the brain to be glamorous, like the movies."

"You've just wanted to be a star ever since *P.S. Your Cat Is Dead,*" I countered. This Broadway show, for which Michael had been the lighting designer, had won the Tony award. The experience instilled in him the hope that his own designs would one day win the Tony. "I think it's likely you'll be seen more as Frankenstein's monster than as Vicki Lester," I suggested, referring to the lead in *A Star Is Born.*

"Oh shoot," smirked Michael, banging the table with his fist.

I shook my head. "You're just like everybody else," I told him. "What most people know of science is what they see on television: men and women putting on white coats and 'doing things.' People have so little first-hand contact, even at the simplest levels. Over half the high schools in the United States, for example, don't offer a single course in physics."

"I had no idea," Michael said.

"It's because of this lack and because chemistry is still taught as if students were going to be bartenders that most people in this country think science is identical to technology."

Michael looked confused. "Well, there *is* some relation," he started to say.

"Engineering is closer to technology than science is," I interrupted. "While people insist on equating science with things, it is far different from technology. Science is an exercise in human imagination. It has different philosophies and points of view. It isn't a monolith."

"The result is still the same, isn't it?" Michael pointed out. "We end up knowing more about how the world works, but we also end up with gadgets in the bargain. Look at Velcro, which we use everywhere in the theater. That came from the Apollo moon landing."

"Gadgets, like Velcro, are what stick in people's minds," I said. "But what did the average person learn from the moon landing?" I asked. "What has filtered down to us in the last ten years?"

Michael was silent. "Too often," I said, "technology evokes a sense of wonder instead of understanding, and I think this makes it a corrosive force which sometimes requires opposition. Those who don't understand it become passive in their amazement, and then they become controlled by it."

"You don't think it's malevolent, do you?" Michael inquired.

I shook my head. "Not yet. I just think it's possible to admire the cleverness of our latest technology without believing that it represents the peak of human achievement." Michael did not look convinced.

"Look, engineering is wonderful," I admitted. "It restructures the world and lets us reach farther with our hands, our voices, and our minds. But technology's social effects, especially the negative ones, always seem not to be anticipated."

"So you're disturbed by the possibility of science being misunderstood, or even misappropriated?" Michael suggested.

"Not precisely," I answered. "But I think that people put too much faith in it, simultaneously becoming more willing to believe anything that invokes the mantle of science, but less capable of critically judging the claim, no matter who is making it. I think the opportunities for manipulation are frightening."

"Doesn't science tell us the truth?" Michael asked. "Isn't that its point, to show us how the world really is?"

"People think so because the values of science don't get taught by a gadget. Science's theme is how things work," I agreed, "but it also has values, especially respect for the use of evidence and logical reasoning. It's also honest, curious, and open to ideas while remaining skeptical when evaluating new claims. The scientific enterprise aims to produce *verifiable* knowledge, that's all. None of the sciences, as far as I know, claims to provide knowledge that is absolutely true. It gives us only one kind of knowing."

And on it went. We had a lively but inconclusive discussion as friends often do. It interested me how Michael in fact had changed from an object of study to a person. Unfortunately, I was unable to tell him any of the results I learned about his synesthesia since it would influence the results. If either he or Victoria knew how they were "supposed" to respond, it would invalidate future experiments. Michael agreed to wear two hats: one as my friend, and another as an experimental subject purposely kept in the dark.

"Let's go back to Frankenstein's monster," he said. "I'm a little worried."

"About looking like a freak when I tell the INS about you?"

"Yeah," he said. "I guess it goes back to feeling silly."

I had explained to him back in North Carolina that what makes science valid is its ability to recognize falsehood. "The hypothesis we're erecting is that you're just like everybody else," I told him, "and we will try to falsify that. If we prove it false, then you have no reason to feel silly."

Michael looked uneasy. "What if it's not false?" he asked.

I hesitated in answering. "If it isn't false, then our works will have been nothing more than an interesting interlude," I said. "And I suppose if you still want to be special, then you'll have to go out and win that Tony award."

▲ ■ ●

My new duties at George Washington kept me busy, as did immersing myself into a fresh medical community and into Washington itself. The year passed quickly and was punctuated by exciting events, not the least of which was the assassination attempt on President Reagan and the brain injury to his press secretary, James Brady. I was appalled at the misconceptions of the public and of his press colleagues.

That summer, I wrote my article for the *New York Times Magazine* and was lucky enough to spend two days with its editor. I had expected glamour but found the *Times* office a dump. I could only imagine that people worked there because of its literary reputation. I quickly reconciled myself that it was *the dump* for aspiring writers, and enjoyed myself immensely.

Lastly, I had also undertaken another big task, which was starting my own private practice. It was a full, rich time in my life. During the course of these events, the wheels of academe turned quietly. My paper had been accepted at the next North American meeting of the

International Neuropsychological Society. I would present it in
Atlanta in February 1981.

▲ ■ ●

"The experience of synesthesia is concrete," I summarized to the
audience. "It is more like a sensation than like an abstract idea.

"By seeing whether the associations are invariant or whether they
are influenced by context, the experimental setup is designed to
choose among the three alternative levels I have sketched in this
diagram," I said, pointing to a slide of my cone.

"Each subject sampled a few stimuli hundreds of times, and we can
see what pattern emerges by plotting how often a certain shape or
color is associated with each stimulus.

"The matching pattern of both synesthetes is consistent with the
predictions of an intermediate level of association," I emphasized,
"while all the controls made the kind of matches predicted if the
associations took place at the high level. The fact that their associations occur at different levels falsifies the hypothesis of no difference
between the perceptions of synesthetes and controls."

Both Michael and Victoria showed the modest amount of contextual variation that the intermediate position predicted, although it
was a matter of quality rather than one of degree. The frequency patterns showed that synesthetic associations were invariant in one part
of the test range, but contextual when the *entire range* was explored!
In response to various combinations of sweet and sour, Michael
showed an invariant response in the sour part of his flavor domain.
Among twenty-three possible shape choices, he used only three to
make all of his matches, all pointed or angular shapes that are conceptually close to one another.

The six control subjects used a variety of strategies in matching
sound-to-color and taste-to-shape. They all spread their responses out
over the available choices, giving very diffuse patterns although not
random ones. Their choices were dominated by context, as would be
expected if they were making them based on abstract qualities. One
control explained that he deliberately decided to use certain shapes
depending on his judgment of whether the taste was mostly sweet or
mostly sour, while another claimed to have no logical scheme. "I just
pick whatever comes into my head." Finally, all the controls
acknowledged making conscious decisions about what to pick and
denied experiencing synesthesia.

In listening to musical notes in a low octave, a high octave, and stretched out over four octaves, Victoria showed the invariant effect when listening to the single, high octave on Tape III. Her pattern suggested that notes she perceived as "high" were pink, while those that seemed "low" tended to be blue. This apparent polarity was the perfect opportunity to explore whether there was any meaning inherent in her associations. I did this with a procedure called the *semantic differential*.[4]

"Some of you may recall that in 1957 Charles Osgood, at the University of Illinois, proposed that *meaning* mediates our mental representations of things. He developed a kind of measurement in which a subject 'differentiates' the meaning of a concept by judging it against a series of contrary scales.

"This slide shows the concept of 'Father' judged against scales of opposite adjectives. You mark it according to how good or bad, fast or slow, or whatever you judge 'Father' to be.

FATHER

Good ____ : X : ____ : ____ : ____ : ____ : ____ Bad
Fast ____ : ____ : ____ : ____ : X : ____ : ____ Slow
Hard ____ : ____ : ____ : ____ : ____ : ____ : X Soft

"Even in people whose language is sophisticated, half the variation in meaningful judgments about any concept is accounted for by just three factors," I explained. "These factors are evaluation, potency, and activity. Evaluation means whether something is considered good or bad. Potency refers to power and to qualities associated with power, like size, weight, or toughness. Activity refers to characteristics such as quickness, excitement, warmth, agitation, and the like.

"What we have learned in the last twenty years is that the semantic differential is a generalizable measure. There are no standard concepts and no standard scales. You make them up depending on what you are studying. While words are used most often as the concepts, the process has been successfully used with Rorschach ink blots, paintings, sculpture, and even sonar signals.

"Precisely because it taps the connotative aspects of meaning more than its denotative aspect," I concluded, "the semantic differential is applicable to æsthetic issues and concepts such as synesthesia."

Victoria differentiated both colors and musical notes over twenty-five scales of judgment. I found no shared meaning between pink and those notes that were perceived to be "high" nor between blue and those notes judged to be "low." In fact, *blue* was judged to be high, good, somewhat passive, and neutral in terms of potency. She judged pink to be neither high nor low, neither good nor bad, neither active nor passive, and only slightly potent.

"Whatever caused her tendency to perceive high notes as pink and low notes as blue," I reasoned, "it was not any shared meaning between the two concepts. On the other hand, the portraiteur who was one of her controls showed context effects in matching red, yellow, and pink to high notes; his semantic differential revealed the meaning of these colors to be all good, potent, passive, and *high*. You can conclude that sensory associations in non-synesthetic persons are mediated by meaning, whereas in synesthetes they are not.

"Recalling Aristotle's common sensibles, it is clear that we can perceive an object as light or heavy by the sound of its fall. We can estimate the number of pins knocked over by the sound of a bowling ball as it crashes into them. These kinds of familiar cross-modal associations might be a mental shorthand that serves to highlight, conveniently, important sensual qualities that different objects might have in common. However, these kinds of attributes are worlds apart from the experience of synesthesia and should not be confused with it," I cautioned.

Like a detective in a drawing room rounding up the suspects, I rounded up my arguments, my evidence, and my conclusions. "In considering all the facts," I told the audience, "I have shown that synesthesia is a sensory experience rather than a flight of imagination. Secondly, it is an experience unlike the cross-modal associations that are the foundation of abstract capabilities such as language, and which are known to occur at a high mental level as well as a high level in the cortex.

"Thirdly, the associations in synesthesia occur at an intermediate level in which the mapping is neither completely one-to-one nor richly one-to-many. The link is mostly invariant, further supporting its location at a low to intermediate level of the nervous system.

"To close on a speculative note, let me suggest that these features also explain why synesthesia is more like a sensation than the kind of abstract idea one gets from an ordinary cross-modal association."

I received polite applause and a few questions from the audience. The session at which I was speaking focused on extraordinary human abilities such as photographic memory and idiot savants. "We've heard a number of unusual topics in this session," a woman commented. "Is it safe to study subjective experience again," she asked, "and what do you think you learn from it?"

I welcomed her question, but could tell from the heads that turned to see who had asked it that the topic was still a touchy one. "I think today's objective versus subjective debate is overblown. Subjective experience was respectable earlier in the century," I pointed out. "We shouldn't be afraid to re-examine our assumptions that human judgment can't be trusted because it is too ambiguous or complex, or that what cannot be measured either does not exist or is irrelevant."

A few people nodded. "Neuroscience should be able to help us understand our subjective experience better," I suggested. "I agree that it is messy to do this. Creative ideas and approaches are needed to meet the challenge of this messiness, otherwise inner experiences will stay beyond our grasp."

"You emphasized synesthesia as a sensation, but it doesn't completely seem like one to me," a red-haired man said.

I thought a moment. "I think many of the lines that separate our mental categories are not drawn as sharply as we assume they are. Take a hallucination: it is like a sensation and like a dream at the same time. Eidetic memory is the same, partly like a perception and partly like a reminiscence. Our categories make the world seem black and white. But it isn't," I answered.

My questioner sat motionless. "Any creative person knows that life is neither black nor white at one time, but changes between extremes," I continued, "How, for example, can nature be both beautiful and ugly, simultaneously creating and destroying?" I groped for an example. "Imagine mature tadpoles eating the younger ones to survive. Is that beautiful or ugly?" I asked. "I think that it is both. In a similar way, humans have many facets and multiple minds."

"Could you speculate further on that?" someone else asked.

"Well," I said, accepting her invitation, "dreaming is the most obvious example of one facet. Where, for example, does the 'I,' the person you think you are, go when you dream? You live a whole other life while dreaming; but you wake up with a sense of continuity to conditions in your waking life, as if your mind never went away.

"Other examples are having an insight, an æsthetic or spiritual experience, crying or being angry for no apparent reason. Such experiences seem to come from beyond events as we know them on the surface, but still they happen all the time.

"The idea of multiple facets hinges on the distinction between the cognitive mind—which is that part that analyzes and demands reasons—and other aspects of our mental life, which could care less about reasons but are interested in the experience of living. Let me give you an analogy between the different facets of a human mind and the duality principle of light.

"The duality principle states that while each photon is an individual particle of light, called a quantum, it is also a continuous wave at the same time. Modern physics has proven that something which is totally individual (a photon) can also be something continuous (a wave). The wave and the particle are both true and valid descriptions of what light is, and an analogy can be made to the human mind, which can also be different things at different times, or even different things at the same time."

▲ ■ ●

The INS affair had brought my work to a close. In the ensuing months I wrote up the details, which were published in the journal *Brain and Cognition*. I was proud, of course, but knew that in the larger scheme of things it was a tiny voice in a screaming chorus, one paper among hundreds of thousands published yearly in the medical literature.

As far as I was concerned, I had left the ivory tower of academe to start my private practice, which I named Capitol Neurology. It was a collection of neck-up specialists in adult and child neurology, ophthalmology, neurosurgery, otolaryngology, and psychology. Since I was a product of my time, we had plenty of technology in the office too.

The days of having time to think about sublime puzzles like Aristotle or synesthesia were over. When faced with people having strokes, tumors, multiple sclerosis, and seizures every day, academic questions seem like crossword puzzles: a mentally stimulating and even amusing pastime, but one irrelevant to the everyday world.

I had left the quiet world of contemplation. I was now part of the system.

SEPTEMBER 1983:
"BIZARRE MEDICAL ODDITY AFFECTS
MILLIONS!"

"I figured this had to be you because it couldn't be nobody else with that unpronounceable name like yours."

The headline horrified me: *"Bizarre Medical Oddity Affects Millions!"*

"Wait till I tell the people at work!" Cecile laughed. "They're not gonna believe my doctor is in the *National Enquirer!*"

Scenes from my life flashed before me as I stared at the headline. *National Enquirer* stories were not known to help one's career. My stomach sank as I thought of my colleagues' likely reactions. "Do you mind if I make a Xerox of this, Cecile?" I asked, my voice rising.

"Hell no," she said. "I bought two copies so you could keep one yourself."

Cecile Bowlding was one of my first patients. Over the ensuing years she would become one of my favorite ones. It is not widely known that doctors do have favorite patients, people who carve out a private spot in our hearts. The reasons they become favorites are as varied as people themselves. Cecile had some wonderful, inarticulable quality; she was also terrified of everything that medicine had to offer, particularly needles, spinal taps, and big machines. She nearly fainted from dread at the thought of having a CT scan. I suppose the thing that made her a favorite was the challenge she posed in providing the best possible doctoring with the least technology.

The "millions" of synesthetes referred to in the headline was tabloid hyperbole. Yet the *Enquirer* story was part of a flood of self-perpetuating media interest that started with a sidebar in *Omni* magazine. A reporter had covered the INS meeting and found synesthesia to be the stuff of snappy headlines. "Azure Words, Mint Triangles" described me explaining that synesthesia was not a disease but "like a bonus. Your senses give you more than you bargained for."

From *Omni*, media interest in synesthesia spread over the next decade, not like wildfire but more like a slow steady glacier. *Psychology Today*, National Public Radio, the *Washington Post Magazine*, Canadian National Radio, radio and television talk shows here and in Canada, and even Voice of America were some of the mainstream parties who interviewed me, sometimes with one of the forty-two individuals whom I eventually studied in detail and who became the basis for my textbook *Synesthesia: A Union of the Senses*.

I failed to anticipate this interest. Lay people were fascinated by synesthesia. It did not seem to threaten their preconceived notions as it did those held by my medical colleagues. In fact, the first public mention of synesthesia in a century triggered impassioned letters from those who, like Michael, never even knew that there was a word for their experiences. The first letter, from an educator in Florida, was typical of what would follow in the ensuing years.

"Dear Dr. Cytowic: I nearly fell over when I saw the article about you in *Omni* magazine. I ran to my husband shouting 'See! This is me! I told you it's real. I'm not nuts!'"

"I feel for the first time like I'm not nuts!" wrote an exultant computer programmer from Arkansas. "When I told people about hearing colors as a child, they looked at me like I needed to be committed. I stopped talking about it a long time ago."

"Unfortunately," wrote an urban planner from Massachusetts, "I only caught the tail end of the TV program, but what I did manage to see absolutely astonished me. I am forty-eight years old and in spite of talking to everyone who would listen, I never came across anyone like me who vividly sees numbers in color and uses the consistency of those colors as a memory aid."

A Canadian surgeon cried when he heard me talk about synesthesia on the radio. "You don't know what a sense of relief it is just knowing that this is real," he wrote.

While the letters from probable synesthetes included anecdotes of their individual associations, as a group they might as well have been writing the same letter. Their stories were remarkably alike, pouring out a mixture of amazement, relief, and emotion—amazement especially that there was a name for something that others had repeatedly denied them as real. To learn that a medical doctor had validated what previously had been a private experience brought about an emotional catharsis, an unburdening and a sense of joy in discovering that something the writer had kept secret for years was shared by other people.

As expected, something as bizarre as synesthesia was also certain to attract its share of kooks and cranks. They too succeeded in tracking me down. The crazy letters were also remarkably alike except that they were dramatic, odd, and revealed the writer's urgent hope to be considered special, as in these three different excerpts:

> In the last six years I have been hospitalized twice for hallucinations and psychotic reactions when under extreme stress. After seeing the TV program about synesthesia, I realized you may have saved me several thousand dollars in therapy!!! . . .

> I have had a lifelong interest in psychic phenomenon. My husband is clairvoyant. He sees a steady stream of pictures before his eyes . . .

> I have a special gift with my hands that I need to develop with supportive people. Blindfolded, a friend places my palm above a plant and I pick up energy feelings. A long leaf will feel like my entire arm is a leaf; a cluster of needles feels like staccato pinpricks in my fingertips. You should know I am recovering from being stressed out most of my life . . .

▲ ■ ●

What was I going to do? I had shut the book on synesthesia, and yet now was being rained on by likely new cases. It was a shame not to do anything. As the numbers mounted, I decided to gather data and study as many of them as I could.

My habits of writing and collecting clinical observations were already ingrained. I had previously written music reviews for the *Winston-Salem Sentinel* and did a brief stint covering restaurants for the *Washington Blade* under the pseudonym Richard Escoffier (the editor said this name sounded "the least made up" of those I pro-

posed). There were also magazine articles about laetrile, food addi-
tives, the piece on Maurice Ravel, another on swine flu, and a brief
medical biography about Anton Chekhov (I had no idea he was a
physician until I saw a performance of *Uncle Vanya*). My first medical commentary, concerning anatomic details of oceanside sunbath-
ers, appeared as a letter in the *New England Journal of Medicine* while
I was still a medical student. Subsequent missives to that journal
concerned taste, emotion, and skipped menstrual periods in women
who had concussions (no kidding).[5] The additional details that I collected regarding people with head injuries later became part of a
textbook.[6]

I imagine that the handful of private practitioners who publish
their clinical research do so because we are more compulsive rather
than bright. Research and publishing is considered the job of university professors who are given time and grant money to do it. Yet I
found that sharing what you have learned requires little of either,
mostly drive and a willingness to deposit observations that are not
yet part of medical lore into the communal literature.

Reading twelve different journals each month gave me a fairly
broad sampling of what the university professors did publish.
Whether the article reported groundbreaking research or hackneyed
drivel, the author nearly always acknowledged grant support. My own
experience with science on a shoestring made me wonder to what use
the money could possibly have been put in some cases.

Patients who travel to great medical centers often have enormous
expectations, and these can be dashed when they discover that some
institutions exist more for the sake of the institution than for ill people such as themselves. That is, not everyone is welcomed with
equally open arms, because the centers want to recruit cases of specific diseases that meet the requirements of their grant-funded
research protocols. Patients' expectations are further disappointed on
hearing that standard phrase: "There is nothing wrong with you."

Yet I had seen that if you listened without interrupting and tried
to look beyond what you "knew" was "impossible," you could often
explain patients' experiences. Even more gratifying was rediscovering, from old books, a treatment for those whose suffering had
sometimes lasted for years. This kind of historical approach yielded
another text, this time about the treatment of chronic pain.[7] My
compulsive work habits and general attitude about medicine produced three consequences: (1) I saw a kind of patient who never got

referred to great medical centers and was therefore unfamiliar to the professors, (2) I investigated patients in greater depth than most private practitioners were likely to do, and (3) I was driven to write down what I encountered.

I applied these old habits to synesthesia as I kept hearing from more and more possible individuals with this rare talent. I decided to make a questionnaire soliciting further details about their experiences and asking correspondents if they would like to be studied. The questions were phrased to reflect the five diagnostic criteria I had formulated and also to help me decide whether these people were really potential synesthetes or were just weird. I became a synesthesia expert by default.

Not everyone who reached me was a would-be synesthete dying to know more. Non-synesthetic individuals wrote to query or comment, most often about the relationship of synesthesia to art. More than a dozen composers of modern and computer-generated music wanted to know the translation algorithm between sight and sound so that their compositions would be "correct." I have already referred to Scriabin as one composer who tried, through his music, to convey his synesthetic colors. In my textbook, I was able to explore how their own synesthesiæ influenced the paintings of David Hockney and the stylistically unique music of Olivier Messiaen.[8]

This correspondence helped round out what I had learned earlier about the history of involuntary synesthesia by making me more aware of various deliberate multi-sensory contrivances that could be lumped under the term "sensory fusion." The origin of colored music, for example, seemed to come from a theory, prevalent in the Renaissance and systematized by the Jesuit music theorist and mathematician Athanasius Kircher (1602–1680), that each musical sound has a necessary and objective correspondence to a certain color. From the eighteenth century onward, various keyboard instruments were adapted into contraptions that could, by pressing a key, simultaneously project colored light and produce a musical sound.

The mysticism inherent in colored music—synesthetic or deliberate—is not unique to it, for the idea of a consonance between sound and color was typical of late post-Wagnerian romanticism. For example, Schönberg, who was not himself synesthetic, experimented with colored music in *Die Glückliche Hand*, a short opera written at nearly the same time as Scriabin's *Prometheus*. Schönberg wished to eliminate any distinction between waking reality and dreaming. His score calls for shifting colors to accompany the music and to mirror the emotions of the characters.

Although not without intrinsic interest, the notion that color and music can be translated into each other rests on a fallacy, as I mentioned in reviewing synesthesia's history. The topic of colored music has received scholarly treatment in Scholes' *Dictionary of Music*. In here, and in other musical references, an undertone of skepticism implies that synesthesia is the musical equivalent of purple prose, the result of "mere" psychologic associations! Perhaps they confused artworks derived from involuntary synesthetic experiences with those that were deliberately contrived. It is for precisely this reason that neurologists should be concerned to show that synesthesia, even though a sensory talent rarer than perfect pitch, is a bona fide object of sensation.

Just as there are deliberately contrived sound-color compositions, so too have painters been moved by music. Georgia O'Keeffe's 1919 picture, *Music—Pink and Blue II*, is such an example from her series of pictures inspired by music. While the distinction that they are inspired rather than synesthetic needs to be emphasized, it is interesting to note, nonetheless, that O'Keeffe was influenced by Kandinsky's *On the Spiritual in Art*. Without having the direct experience of synesthesia herself, perhaps she still captured in a more circuitous way the noëtic quality of her subject that cannot be put into words. Her discovery that colors could carry psychological and emotional states of mind and her convictions about the expressive power of abstract art are clearly stated in a letter from 1930.

> I know I cannot paint a flower. I cannot paint the sun on the desert on a bright Summer morning but maybe in terms of paint color I convey to you my experience of the flower or the experience that makes the flower of significance to me at that particular time.[9]

Here may be an example of the directness of perception characteristic of both synesthesia and artistic vision. Both are ineffable, both truly indescribable. It was examples such as these that prompted me to think that synesthesia was not an isolated, esoteric quirk, but that some aspect of it pointed to a noëtic understanding of which everyone is capable. By this I mean your ability to understand something directly without knowing how you understand it.

The inspiration of synesthesia popped up in many places. For example, the Hungarian composer Zoltan Kodály (1882–1967) invented a method of teaching music to deaf students by the use of hand signs, each hand position representing a note value. Popularly, Kodály's system and the idea of joining hearing to movement was demonstrated in the film *Close Encounters of the Third Kind,* in which an alien spaceship visits Earth. The alien "message" to Earth is a melody accompanied by colored lights that emanate from the ship. The earthlings respond by parroting the melody and colors even though they have no idea what they are "saying." At last, a wise scientist uses Kodály's formula for sound and hand motions, and deciphers the message as a gesture of greeting and handshake.

In scientific circles, the idea of a unity of the senses may have taken root in an obscure 1926 article by the German psychologist Von Hornbostel.[10] He described a sensuous state called a supersensuous sense perception, the essential component of which unites all the senses "among themselves, unites them with the entire (even with the nonsensuous) experience in ourselves, and with all the external world that there is to be experienced!" How much the "objective" scientist sounds like a symbolist poet!

Of all those who wrote to me, the most charming inquirers were elementary and high-school students eager to write book reports and devise science fair demonstrations about synesthesia. Even my nephew was part of this group. They were excited to read about my work but dismayed in coming up empty-handed at their local library. I happily complied with their urgent requests for help with their school reports.

As the years went by even mainstay institutions began to show interest in my work, a sign that professional resistance to subjectivity was lessening. Four graduate students proposed writing their Ph.D. dissertations on synesthesia and requested advice. The National Science Foundation and the American Association for the Advancement of Science both invited me to give a lecture, and the *Encyclopedia Britannica* commissioned an article. It looked as if synesthesia and I were in danger of becoming respectable after all.

Chapter 15

▲ ■ ●

FORM CONSTANTS AND
EXPLAINING INEFFABLE EXPERIENCES

Over the next nine years, the mailman continued to bring me surprises. Total strangers expended great effort trying to explain what their synesthesia was like. No matter whether they sent five-page letters, sketches, or paintings that tried to capture their sensations, they uniformly apologized that the materials could not even begin to convey what their experiences were "really like." What the mailman brought, I understood, were representations and not reproductions of synesthetic experiences. A woman named Rachel wrote about its indescribable nature:

> Dear Dr. Cytowic: I read the article concerning your work with synesthesia. You have no idea how exciting it is to read someone else's description, from a total stranger, of an experience that I have never been quite sure wasn't the result of my imagination or being insane.
>
> I most often see sound as colors, with a certain sense of pressure on my skin. I have never met anyone else who saw sound. I'm not sure that "seeing" is the most accurate description. I am seeing, but not with my eyes, if that makes sense. I can't imagine being without my colors. One of the things I love about my husband are the colors of his voice and his laugh. It's a wonderful golden brown, with a flavor of crisp, buttery toast, which sounds very odd, I know, but it is very real.

Her comment that "I am seeing, but not with my eyes," was important. Even when synesthesiæ are experienced outside of one's body, the experience is otherworldly. "Seeing" is not done wholly with either the eyes or the mind; perhaps it is best to say that it is done partly with both. Of course, synesthetes can picture things in their imagination just like anyone else; yet they insist that their synesthetic experiences are nothing like normal imagination. It is so difficult to describe this sense of inhabiting two worlds at once, like being half awake yet still anchored in a dream.

These people were trying to explain the ineffable, that which by definition cannot be put into words. My favorite definition of "ineffable" as a private experience was given by the pioneering American psychologist William James in his 1901 book, *The Varieties of Religious Experience*. "The subject says that it defies expression, that no adequate report of its content can be given in words. It follows from this that its 'quality' must be directly experienced, it cannot be imparted or transferred to others."

Synesthesia is exactly like this. The struggle to explain the experience seems as impossible as trying to explain what hearing is like to someone born deaf. The lack of a shared referent is only part of the problem, because any experience really can be told if you choose your words carefully. Telling, of course, does not guarantee understanding. Compounding the problem is a bias, among Western institutions, against examining inner knowledge and higher creativity coupled with an authoritarian tendency to define what is "normal."

By the time they are adults synesthetes have usually encountered so much disbelief that you have to coax them to talk. "Nobody understands," "People look at me like I'm crazy," or, "I don't want to be thought of as a freak," they say. Together with the inherent difficulty of conveying the ineffable, this climate makes synesthetes reticent to share their experiences. Crackpots, on the other hand, will talk about their "visions," "mental powers," and "psychic vibrations" at the drop of a hat.

The only remarkable exception to this reticence is when synesthesia runs in the family. Seven of the forty-two individuals I eventually studied had immediate relatives who were also synesthetic. In one woman's family, I was able to trace the trait through four generations.[11] I was also delighted to discover that the writer Vladimir Nabokov discussed his colored hearing, and that of his mother, in his novel *Speak, Memory*.[12]

I was impressed by how highly individualized the triggering stimuli usually are, a fact that might explain why the expression of synesthesia varied so much from person to person. That is, while it is an all-

120 THE MAN WHO TASTED SHAPES

or-nothing trait (you either have it or you don't), some people seemed to "have it" more than others. Another woman named Anne, for example, saw colored shapes only in response to a specific type of music, and not any other kinds of sound.

> I see shiny white isosceles triangles, like shards of broken glass. Blue is a sharper color and has lines and angles. Green has curves. I feel the space above my eyes is a big screen where this scene is playing.

Like the triangles, lines, and curves in Anne's example, what synesthetes actually sense is elementary. Their reactions only seem more elaborate when words fail them and individuals resort to analogy to convey their sensations. They say the experience is "like" something else which is familiar. The uncritical mind is liable to leap to the false conclusion that the parallel sensations represent mere imagination. When we say that red is a "warm" color or that a certain cheese tastes "sharp," we speak metaphorically. No one pretends to have a thermal sensation to the color or a tactile one to the cheese. With synesthesia, however, a comment like "that's really a sharp color on you" could be literal.

The opposite extreme for people who, like Anne, have very limited triggers as well as having their synesthesia limited to one sense is someone in whom everything is connected. Stimulation of any one sense causes synesthesia in the remaining four. For example:

> I heard the bell ringing . . . a small round object rolled before my eyes . . . my fingers sensed something rough like a rope . . . I experienced a taste of salt water . . . and something white.[13]

Such a world might seem like a nightmare to those of us whose senses keep to themselves. How, we wonder, can anybody think with all this extra stuff flying about in their brains? Why don't they go insane from sensory overload? Possibly, the stability and constancy of synesthesiæ keeps them from seeming chaotic or confusing. Empirically, synesthetes appear normal and quite intelligent; they seem to be in little danger of going insane from all the "extra stuff" of their experience.

Our extreme difficulty in comprehending synesthetic experience and our presumption of what is normal leads to the false assumption that synesthesia must be a burden or somehow interfere with normal thinking. By analogy, blind persons would be justified asking why sighted individuals are not bothered by having to constantly "see everything." Are we not driven to distraction by a constantly changing visual jumble? It seems that we are all able to handle our own subjectivity and make sense out of our experience, even though it may be quite different from someone else's subjective experience.

It is easier to approach the synesthete's world if we let go of judgments and preconceived notions about how things are supposed to be. For all the talk about seeing, feeling, and tasting fantastic things, careful observation shows that *what synesthetes actually sense is always quite simple.* For example, they *never* see a landscape while listening to Beethoven or feel an oriental lacquered box when eating clam chowder. Rather, they see blobs, grids, cross hatchings, and geometric shapes; feel rough, smooth, or prickly textures; taste salty or metallic tastes. All of these are generic perceptions. It may be that the simple elements of synesthesia are really the building blocks of more complex human perceptions.

The problem here is twofold. The first deals with synesthesia's ineffable quality. In our limited ability to understand what they sense, our own imaginations jump in to fill the gaps. Yes, we think, we might imagine a beautiful landscape while listening to Beethoven. But what we *imagine* might happen and what synesthetes *actually experience* are different things. Synesthetes cannot help what happens to them; we can only fantasize what might happen to us.

Secondly, the only way for synesthetes to share their perceptions is by telling us about them, and examples from neurology have already shown that nothing can be taken at face value. Michael Watson, for example, described the taste of mint as "cool glass columns." But what does he really mean by this? His description might lead you to assume that he is being poetic or using metaphor. He is not. He is giving a *verbal interpretation of a sensory experience.* What if we asked him to cut the interpretation and give it to us as raw as possible?

One evening I pressed him to describe the qualities of touch that he felt and to explain how he reached his interpretation of a glass column. After taking a sample, he closed his eyes and paused. His right hand swept vertically through the air as he moaned pleasurably.

He rubbed his fingertips together and moved his hand through the air as if palpating an invisible object.

> I feel a round shape. There's a curve behind which I can reach, and it's very, very smooth. So it must be made of marble or glass, because what I'm feeling is this satiny smoothness. There are no ripples, no little surface indentations, so it must be glass, because if it were marble, I would be able to feel the roughness of the stone, the pits in the surface. It's also cold so it has to be some sort of glass or stone because of its temperature. What is so wonderful is the absolute smoothness of it. I can run my hand up and down, but I can't feel where the top ends. It must go on up forever. So the only thing I can explain this feeling as is that it's like a tall, smooth column made of glass. In fact, with the amyl nitrite [a drug that intensifies the synesthesia], it's as if there's a whole row of them and I can stick my hand in among the columns and feel the back sides of the curves. There is this funny sort of feeling of being able to reach my hand into this area. It's very, very pleasant.

Michael's verbal description of a sensation is like using analogy to explain to a blind person what it is like to see. In my conversations with Dr. Wood, I had suggested that synesthesia was not a special instance of language but rather the opposite. Language would probably never have evolved without humans first being able to form the kinds of cross-modal associations present in synesthesia. This assertion goes back to the discussion we had of cross-modal associations in monkeys, who are unable to associate two non-limbic senses. Humans can do this, and it is this capacity that underlies the ability to assign names to objects and proceed to higher and higher levels of mental abstraction.

FORM CONSTANTS

The indescribable nature of subjective experience is not unique to synesthesia. Heinrich Klüver faced the same difficulty when he tried to understand the experience of hallucinations at the University of Chicago starting in the 1920s. Klüver was frustrated by the vagueness with which subjects described their experiences. He felt they were

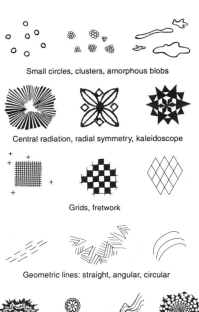

Small circles, clusters, amorphous blobs

Central radiation, radial symmetry, kaleidoscope

Grids, fretwork

Geometric lines: straight, angular, circular

Scintillation, extrusion   Iteration   Movement   Rotation, spiraling

Figure 7

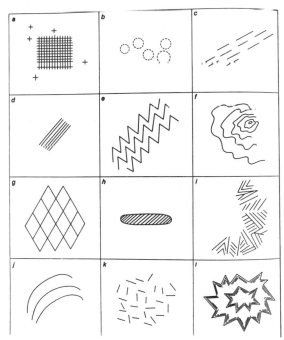

Figure 8

Figure 9

Figure 7. Generic shapes (Klüver's "form constants") are common to synesthesia, hallucinations, migraine auras, and imagery, and can also be seen in primitive art. From MJ Horowitz 1983. Image Formation and Psychotherapy, p. 200. Hillsdale, NJ: Aronson. Reprinted with permission.

Figure 8. Projected visual images, in the eyes-open state, induced by cocaine, show the typical grid, lattice, and linear configurations of the form constants. Axial and radial symmetry are common. From "Hallucinations" by RK Siegel. Copyright 1977 by Scientific American. All rights reserved.

Figure 9. Top: Tunnel form constant, showing a "bright light" with an explosion from the center that radiates to the periphery. Colors may change, and rotation or other movement may be perceived. Bottom: Combination of the tunnel and spiral form constants, with rotation and pulsation. Drawn by the patient from a drug-induced hallucination. From Hallucinations: Behaviour, Experience, and Theory pp. 116–117 by RK Siegel & LJ West, 1975, New York: John Wiley & Sons. Copyright 1975 by RK Siegel & LJ West.

overwhelmed and awed by the "indescribable" nature of their visions, and yielded uncritically to cosmic or religious explanations instead of taking pains to accurately describe what happened. Already in synesthesia we have seen a similar tendency for elementary sensations to be metaphorically elaborated when subjects are sloppy with their descriptions. Only by insisting on a description of their actual sensations can you thwart poetic embroidering.

Once Klüver pinned his subjects down and got them to describe without elaborating or interpreting, he saw that basic patterns of perception existed. Klüver called these patterns *form constants* and identified four types of constant hallucinogenic images: (1) gratings and honeycombs, (2) cobwebs, (3) tunnels and cones, and (4) spirals. Variations in color, brightness, movement, symmetry, and replication provide finer gradation of the subjective experience. He suggested that the novelty of hallucinations and their vivid coloration captivated people's attention and invited unwarranted elaboration of what were basically *elementary features that the nervous system was hardwired to perceive.* The point is that, given an infinite variety of stimulation, the brain seems capable of perceiving in finite ways.

In time, neuroscience came to understand that the brain actually has evolved to perceive fundamental patterns and has built-in filters that recognize constant, possibly useful, features in the energy flux that relentlessly bombards us. It is possible that marked consistencies in the connections of the limbic brain, which are retained throughout its evolutionary elaborations in all living vertebrates, are responsible for the generic and constant features perceived in synesthesia and the other altered states of consciousness that I discuss in the next chapter.

Figures 7 through 9 show examples of Klüver's form constants. In addition to form, there are also color and movement constants, such as pulsation, flicker, drift, rotation, and perspectives of advance-recede relative to the viewer. Form constants can be found in many natural phenomena, from subjective experiences to works of art, including craft work and cave paintings of primitive cultures.[14]

With time the idea of form constants gained acceptance and Klüver's work was repeated and extended by others.[15] What Klüver showed was that there are a limited number of perceptual frameworks that appear to be built into the nervous system and that are probably part of our genetic endowment.

The analysis . . . has yielded a number of forms and form elements which must be considered typical for mescal visions. No matter how strong the inter- and intra-individual differences may be, the records are remarkably uniform as to the appearance of the above described forms and configurations. We may call them form-constants, implying that a certain number of them appear in almost all mescal visions and that many "atyp-ical" visions are upon close examination nothing but variations of these form-constants.[16]

In other words, the redundant elements of generic images and other sensations indicate that "certain consistencies" of the nervous system themselves contribute to illusions and hallucinations, as well as to ordinary perception. These generic perceptions are *identical* in synes-thesia and such altered states of consciousness as sensory deprivation, migraine, temporal lobe epilepsy, drug-induced states, psychosis, and the delirium of fever. For example, Kandinsky described them during a feverish delirium: "Pictures, microscopic preparations, or ornamen-tal figures were drawn on the dark ground of the visual field."[17]

In 1975, researchers at the University of California[18] actually trained volunteers to recognize the form, color, and movement con-stants of drug-induced hallucinations by training them with a series of color slides depicting variations of the constants. Once they had learned the categories, they were able to describe the hallucinations in standard terminology instead of being overwhelmed and awed by the novelty of the experience or resorting to vague platitudes.

The human tendency to attach supernatural meaning to unusual experiences stems from the referential nature of the perception and its emotional impact. Despite human fondness for magical thinking, it is a presumption to attach cosmic meanings to these images. "We are more likely confronting here the projection of affect onto the outside world," says one researcher. "The response that one thinks he recognizes is his own projected and reflected image."[19]

Chapter 16

▲ ■ ●

ALTERED STATES
OF CONSCIOUSNESS

To understand something unfamiliar it is often helpful to look at a familiar thing that we do understand. This is called reasoning by analogy, and is related to the second mainstay of scientific investigation, description. Synesthesia happens to resemble a number of other experiences. For example, it is partly like a sensation and partly like a memory, without being exactly like either one. Because we know a lot more about the workings of memory than we do about synesthesia, a look at photographic memory, for instance, might help us zero in on where the synesthetic link is physically located in the brain.

Six altered states that bear some similarity to synesthesia and are much more familiar to neurology are: (1) LSD-induced synesthesia, (2) photographic memory, (3) sensory deprivation, (4) temporal lobe epilepsy, (5) release hallucinations, and (6) direct electrical stimulation of brain cortex. I'll explain what all these states are in just a moment.

Examining these states of consciousness will show that synesthesia depends only on the left hemisphere, that a structure called the limbic system is essential for its expression, and most surprising of all, that it does not rely on the brain's *cortex*. By examining instances of synesthesia that have a known cause, I hope to explain the synesthesia that occurs naturally in persons such as Michael Watson, who have *no* brain pathology at all.

LSD-INDUCED SYNESTHESIA

LSD *sometimes* produces synesthesia in persons who are not otherwise synesthetic. It is not an inevitable effect of the drug, as many people assume from impressions left by hippie counterculture of past decades. Even if LSD produces synesthesia once, there is no guarantee that it will do so every time one takes it.

Ethical considerations guarantee that 1950s-era government research into the effects of LSD on humans will never be repeated. While no contemporary research exists, however, the older data about the drug's general effects on the nervous system are reliable. The subjective experience of volunteers taking LSD is similar to that of natural synesthetes.[1] Both report vivid and concrete sensations that are "not really there," both find the experience emotionally significant, and both retain a vivid memory of it. Additionally, both natural synesthetes and those on LSD experience a small number of perceptions, quite unlike the volunteers in studies of imagination, who typically produce reams of associations in a stream of consciousness.

Electrodes inserted deep into brain tissue reveal a pattern of activation during LSD exposure that is altered from the normal flow. I will go into the details in Chapter 19. For now, you need only know that LSD exerts three physiological actions, two of which oppose each other. First, LSD *enhances* the low-level synapses coming from the sense organs to a brainstem relay called the thalamus, while at the same time *suppressing* the synaptic connections between that relay station and higher brain areas down the line. Simultaneously, LSD causes an overall alertness and a *specific enhancement of synaptic pathways to the limbic system*, the part of the brain that gives salience (relevance) to events and is intimately concerned with emotion and memory. These three physiological events are simultaneous with the volunteer's subjective experience of altered perception.

What this suggests is that the limbic system is stimulated while the cortex, whose function it is to analyze and make fine distinctions, is suppressed. This change in the normal balance of neural sensory flow results in an individual who cannot discriminate but who is prepared to respond emotionally to sensations that are felt more intensely than usual. By blocking the normal flow of nervous impulses at a point before a unified experience is created, LSD makes it "stick" at a detail of the perception, like a stuck phonograph needle, and this is what dominates the subjective experience. For example, a person tripping on acid might look at a rose and later recount how the "whole room" was filled with a huge rose petal. Or the color purple might imbue not only sight but also sound and taste during the hallucination episode.

What this says in relation to synesthesia is that isolated bits of perception (such as color, movement, shape, direction, number, size, and so forth) can detach from the sensory stream and be experienced by themselves. How such a detached aspect can attach itself to another sense with which it is not normally associated is a separate issue.

PHOTOGRAPHIC MEMORY

Synesthetes have terrific memories.[2] Michael Watson volunteered a specific childhood memory about the way the sunlight hit some daffodils. His memory of the "angle" of the sunlight is a highly crystallized one of a perceptual event that occurred on an annual basis. He would observe daily until the luminance was "just so," which it was for only a brief period.

> On a specific morning in April, I could plot it year to year, there was a particular way the sunlight hit on our driveway on the daffodils. It was very vivid and I would look forward to it every year at the same time of day to be exactly the same way. It was the beginning of Spring for me. I knew it was here. I loved it. It was wonderful. It was not going to be cold anymore. I couldn't wait until that day when the light arrived.

In addition to excellent memory for such details, synesthetes have a nearly indelible recollection of the synesthetic sense itself. For example, seven years after Michael's comment about his chicken not having "enough points," the two of us happened to be dining on roast chicken again. I pointed out the irony and misquoted him, saying that there were too many corners. In the process of correcting me, Michael claimed to remember the original stimulus rather than just the anecdote, as I had.

> I recall the taste, particularly the shape. I corrected you by saying, "It's too round and needs points," because that's what it

was, uniformly round. I remember the *shape*, not the anecdote. I remember being disappointed with the chicken. I tasted it and thought, "I can't serve this." I had to give it points.

The vividness of such memories is reminiscent of eidetic memory, popularly known as "photographic memory" because the recollection is so minutely accurate and indelible that we metaphorically liken it to a photographic print. Like synesthesia, the eidetic image is seen externally on a "screen" and is so stable that one can accurately recall details years after the initial exposure.[3] Electroencephalograms show that the cortex of eidetic individuals is suppressed during detailed recollection. This and other evidence points in the same direction that LSD did: away from the cortex and toward the limbic system.

Since 1953 we have known that parts of the limbic system are essential to forming new memories. We know this from the unfortunate amnesia of a noted patient called HM,[4] who had these parts removed in an attempt to relieve his severe epilepsy. Although his epileptic condition markedly improved, the loss of these limbic structures caused HM to forget the episodes of his daily life as quickly as they occurred. No event after his surgery has ever been encoded as a permanent memory, nor has he ever learned who the persons are who have cared for him all these years. Having aged since his surgery, he cannot now recognize a current picture of himself. He literally lives in 1950, unable to assimilate the flow of current events.

Hypermnesis—literally meaning "elevated memory"—is the opposite of HM's situation. Perhaps an *enhancement* of these limbic structures might account for the detailed recollection in photographic memory, LSD adventures, and synesthesia. The cortex and limbic systems have reciprocal connections, and one way to enhance limbic function is to suppress the cortex. That the relative balance of limbic and cortical strengths can be altered is suggested by the experience of Luria's patient S and the few others whose synesthesia overwhelms their thinking. S, of course, had an astonishing photographic memory, but it was his inability to suppress his synesthetic percepts that was often so severe as to make it difficult for him to understand the meaningful qualities of a conversation.

What first strikes me is the *color* of someone's voice. . . . He has a crumbly, yellow voice, like a flame with protruding fibers.

> Sometimes I get so interested in the voice, I can't understand what's being said. . . . Should another person's voice break in, blurs appear. These creep into the syllables of the words and I can't make out what is being said.[5]

S often became trapped in a synesthetic tangle of sensation that pulled him away from the subject at hand. "Static," "puffs of steam," or "a bitter taste" overwhelmed him, made it impossible to comprehend, and led to irrelevant digressions. His concrete images and sensations would guide his thinking instead of thought itself being the dominant element. Luria found it hopeless asking him to convert particular encounters into general instances, a process by which we ordinarily form general concepts even though the particulars on which they are based are soon forgotten.

Perhaps S had trouble with inductive reasoning because the details of any experience are essentially unrepeatable, constituting a singular episode in one's life. The semantic abstractions of a given experience on which generalizations are based, however, are the currency of language and can easily recur during subsequent events. Therefore, it is precisely the *concrete* level of encoding, one which is conceptually meagre but sensually rich, that appears to facilitate the vivid and long-lasting memory for discrete episodes. It happens that the intermediate level that I have proposed as a synesthetic link also possesses this quality of concreteness.

RELEASE HALLUCINATIONS

Let me take the example of vision. The old view of brain organization conceived of vision as a linear process in which shape, size, color, contrast, location in space, and other features were successively added like links to a chain. We now believe that such details are derived simultaneously. That is, the job of determining the shape of what we are looking at goes off in one direction, the job of finding out what color it is gets farmed out elsewhere, and so forth. The map of each detail is different, and while each is derived in parallel rather than in sequence, they are not all processed at the same instant. Therefore, there is a very brief but finite amount of time during which the component features of an experience are assembled.

Errors can occur in this assembly. Blocking the flow of neural impulses early in the chain, at the primary cortex, results in simple blindness, deafness, and so on. Blocking it further downstream, however, causes some of the maps to be processed while others are not. In other words, you do have an experience, but it is incomplete. This is what happens in release hallucinations.

When the primary cortical area belonging to any sense is damaged, another cortical area down the line is "released" from its influence and can generate signals of its own without having received any information at all from the outside world. In doing so, it causes perceptions in the blind, deaf, or numb part. For example, a person with a stroke in the first cortical relay station for vision is blind. Such a patient sees only blackness in that part of the visual field belonging to the visual cortex that suffered the stroke.

Because the downline association cortex is released and unconstrained in sending impulses of its own maps downstream, one such patient saw things in the "blind" part of her visual field. The events that trigger release hallucinations can be astoundingly specific, just as they are for synesthesia. Whenever this woman read a book or watched television, she saw "four or five men moving about, some in business suits, one in a cowboy's suit, one in a plaid shirt." They would abruptly disappear when she stopped these activities, only to return whenever she resumed.[6]

As damage moves away from the first cortical relay and toward the limbic system, the released hallucinations become less specific and detailed, and more elementary. In fact, they begin to look like form constants. One patient with such a lesion (who was not synesthetic) experienced three types of hallucinations. These were red and green perpendicular lines that moved toward him; stationary red and blue dots; and black and white pulsations.

As we saw with LSD, evidence mounts that an individual map, which represents a *single quality of sensation*, can be perceived all by itself. Furthermore, people with release hallucinations show us that large amounts of "sensory cortex" are not necessary to have sensory experiences! Like what we have learned about the gustofacial reflex, this is again counter-intuitive, because the old view has ingrained in us the idea that the conscious perception of experience takes place in the cortex. Clearly, we are going to have to rethink this idea.

SENSORY DEPRIVATION AND SIMPLE SYNESTHESIA

The physiology of release hallucinations is perhaps easier to understand if one considers sensory deprivation. Normally, only some of the sensory input that constantly bombards the brain is relevant. Most of it is filtered out. Experience with sensory deprivation—as in John Lilly's famous tranquility tank—shows that removing all sensory input leads to psychotic thinking, perceptual distortions, and hallucinations. Milder degrees of sensory loss (such as cataracts, hearing loss, or the feebleness of touch called peripheral neuropathy) lead to less florid results. But the rule of thumb is that a brain deprived of external input will start projecting an external reality of its own, readily perceiving things that are "not really there." This release of the senses, particularly sight and sound, is not as rare as you might think: when your hearing is drowned out by the white noise of the shower, for example, how often have you hallucinated that the phone was ringing or that someone was calling your name?

Normal persons deprived of sensation progress from having mild to severe hallucinations, starting out with what looks very much like form constants (geometric patterns, mosaics, lines, rows of dots) and building to more developed, dream-like juxtapositions of perceptions the longer they remain in isolation.

Auditory-visual synesthesia is not uncommon in patients who injure their optic nerves, yet its occurrence is rarely appreciated because of the dismissive attitudes and pinched views of ophthalmologists. Besides, the experience is temporary and disappears shortly after the onset of blindness. One exceptionally attentive physician investigated nine of his patients who had sound-induced photisms.[7] Such sounds as the clanking of a radiator, the crackling of walls as they cooled at night, the whoosh of a furnace ignition, or a dog's bark caused them to see photisms ranging from simple flashes of light to colored forms that looked like a flame, blobs, pulsating flower petals, a spray of bright dots, or a kaleidoscope. These patients were otherwise normal, sound of mind, and appreciated the unreality of their unusual experience. A similar occurrence of musical and vocal hallucinations, disturbing only because of their monotony, was documented in a woman with progressive hearing loss.[8] The opinion that reduced sensory input can lead to hallucinations in sane people was first proposed in 1894, an opinion that has not changed in the past century.[9]

Our reasoning by analogy in these examples leads to the same conclusion we made when we theorized about the possible level of the synesthetic link. Namely, the level at which synesthesia operates must exist above the level of nuclear relays in the thalamus but below the primary sensory cortex, since the cortex in these patients functions normally. If we think about the flow of impulses as linear, we get trapped because the old view of brain organization places the thalamus and primary cortex next to each other. That is, there are no levels in between! The contemporary view thinks in terms of parallel processing and resolves the issue of levels by pointing to a structure to which all sensory impulses branch. That structure is the limbic system.

Our thoughts about this possibility are greatly helped by a patient with auditory-visual synesthesia caused by a tumor in the temporal lobe of his left hemisphere and whose synesthesia stopped after removal of that tumor.[10] This case not only suggests that synesthesia may be asymmetrically represented in the brain, but also points to that part of the limbic system which resides within the temporal lobe.

TEMPORAL LOBE EPILEPSY

An inherent property of a nerve cell is the repeated building up and discharging of its electrical energy. A seizure is a sudden, simultaneous discharge of many nerve cells. Where this sudden and synchronous discharge begins, how long it lasts, and whether it spreads (perhaps to the entire brain itself, as in a *grand mal* convulsion) determines its clinical manifestations. Over a dozen clinical seizure types exist. Some have purely physical manifestations, while others have mental, sensory, or emotional ones either singly or in combination.

Components of the limbic system within the temporal lobe happen to have a very low threshold for epilepsy, so that seizures can remain confined within the limbic system and produce both psychic and motor manifestations without spilling over to the rest of the brain. (The terms temporal lobe epilepsy, psychomotor epilepsy, and limbic epilepsy are used synonymously.) *The most distinctive characteristic of temporal lobe epilepsy (TLE) is a qualitative alteration of consciousness.* Relative to my notion that humans possess multiple minds, TLE demonstrates well that a level of mentation exists that is not accessible to awareness but that indisputably exerts observable effects on both behavior and subjective experience.

The motor effects of TLE include automatisms, in which the patient performs a variety of well-coordinated actions that seem rational and purposeful to an uninformed observer but that the individual does not recall. These actions are performed without any awareness whatsoever. Sensory and psychic manifestations of TLE include hallucinations, especially of smell and taste; perceptual distortions such as out-of-body and the inverted- and reversed-visual experiences mentioned earlier; and subjective experiences such as *déjà vu* and *jamais vu*, and overwhelming feelings of religious bliss, clarity of mind, or certitude (the "this is it" feeling).

In respect to synesthesia, what interests us is that the discharges of TLE can join the elements of smell, taste, vision, touch, hearing, memory, and emotion. Epileptic synesthesia occurs in four percent of temporal lobe seizures. Epileptic synesthesia ranges from well-developed experience (seeing "beautiful places, large rooms" and hearing "beautiful music" at the same time[11]) to elementary sensations such as "heat," "a taste," or "numbness." These examples of epileptic synesthesia were the sole manifestations of the actual seizure in four different patients:[12]

- A taste of bile, numbness of the left wrist, twitching of the left corner of the mouth
- Stomach pain, shivers, a bitter taste, nausea
- A lump in the throat, mouth and tongue movements, photisms up and to the right, a bitter taste
- An intense heat rising from the stomach to the mouth, accompanied by a disagreeable taste

Analysis of such cases with electrodes implanted in the brain, plus surgery at times, shows that taste and smell sensations arise when the seizure involves limbic structures called the amygdala (a nucleus) and hippocampus (a simple type of cortex with only three layers). Tastes are never described in detail but in general terms such as "bitter," "unpleasant," or just "a taste." The analogy with the generic sensory qualities of synesthesia is apparent. Only when the seizure spills over to the more elaborate six-layered association cortex of the temporal lobe is the taste perceived more specifically ("rusty iron," "oysters," or "an artichoke").

Whether a simple sensation or an elaborate experience is produced depends, therefore, on which part of the temporal lobe is stimulated. This distinction that the ancient hippocampus produces generic experiences, such as those seen in naturally occurring synesthesia, while the younger temporal cortex yields more elaborate ones has been demonstrated by other methods, including direct electrical stimulation of the brain.

ELECTRICAL STIMULATION OF THE BRAIN

More than anyone else, the Canadian physician Wilder Penfield, in the 1950s, showed that electrical stimulation of the temporal lobe could make patients relive the past as if it were the present. This was Proust on the operating table, an electrical *recherche du temps perdu*, yet obviously not *perdu*. The evoked memory, or "experiential response" as Penfield called it, went beyond ordinary figurative memory. It seemed like a full, physical re-enactment of an already lived experience. The reminiscence produced when the electrode touched the brain was dynamic, unfolded over time, and was completely out of the patient's control as long as the current stayed on.

Patients were startled to relive conversations, family meals, schoolrooms, neighbors long forgotten, songs they had not heard in years, moments of joy or embarrassment. Like synesthetes, patients who had a temporal lobe stimulated could appreciate "both worlds." They had a strong conviction that what they experienced was real without losing sight of the fact that they were on an operating table in Montreal. As one patient put it, "I see the people in this world and in that world, too, at the same time."[13]

Penfield's work was the first to suggest that memory was stored in different ways, although he did not recognize this himself. Nonspecific generalizations of discrete life experiences that are predominantly intellectual and unemotional are most commonly stored as verbal representations. This is what we call semantic memory. What Penfield and those who followed in his footsteps showed was that it is also possible to recreate the original episode with every bit of its vivid sensory and emotional trappings. Only limbic stimulation evoked experiential responses.

Summary

Five of these altered states of consciousness are similar to synesthesia, while TLE almost exactly replicates it. However, they are all pathological states with an identifiable cause, whereas the synesthesia we are interested in satisfies neither condition. Nonetheless, those instances with an identifiable cause might guide us to the brain mechanism of the idiopathic (naturally occurring) kind. They also force a clarification of terminology. *Induced synesthesia* appears in patients with acquired brain injury, as in the example of the clanking radiator inducing colored photisms in a person with visual damage. That caused by LSD we can call *drug induced*. The term *synesthesia*, without qualification, refers to a lifelong characteristic in those without any nervous system pathology and is the subject of our continued investigation.

What these six experiential states have in common is a disruption or suppression of more abstract processes in the brain's cortex along with a simultaneous blocking of sensory input. They all intimate something special about the temporal lobe and that part of the limbic system within it for generating altered states of consciousness. It is here that I continued my search for the seat of synesthesia.

Chapter 17

▲ ■ ●

MAY 21, 1981: TAKING DRUGS

Sometimes it is difficult for a scientist to leave things alone, especially when it comes to proving a given idea. Since it is falsifiability that makes something scientific, practitioners of science grow more comfortable as multiple lines of evidence support a pet hypothesis. Because a single negative instance can shoot down an idea, no amount of support is ever absolutely adequate; but you eventually reach a point where proof means that enough accumulated positive evidence has made the likelihood of falsification remote. So far, I had proved synesthesia on two levels, the theoretical and the experimental. Now I was toying with the idea of a technical proof based on manipulating neurotransmitters in Michael's brain.

Even though I had left North Carolina I still kept in touch with colleagues. Dr. Wood joined me for dinner one evening in Washington. Besides a stimulating discussion, I got a free floor show watching him devour a bloody slab of prime rib.

"You know, the idea that the brain's cortex is 'turned down' while the limbic brain is 'turned up' during synesthetic experience is suggested by several of the things we've already done," I reminded him.

"What did you have in mind, Rick?"

"I was thinking that changing the relative strength of the cortex compared to the limbic brain would alter the intensity of a synesthetic experience. Drugs are a plausible tool for doing this." I reminded him how increased limbic activity was the likely explanation for hypermnesis and LSD-induced synesthesia.

Dr. Wood spoke from the side of his full mouth. "LSD is unobtainable, and most of the other drugs that could do what you suggest are investigational." He washed down the bolus with three swigs of beer. "Wouldn't you need special clearance from the Drug Enforcement Administration?"

"I was thinking of something much simpler than designer molecules," I told him. "In fact, Michael's own daily routine suggested it."

Michael's synesthesiæ seemed to follow a daily rhythm, being more vivid in the evening compared to the morning hours. The diary I had instructed him to keep showed that he actually experienced synesthesia at all hours, except that it was less developed during the day. "It's not as intense," he explained. "There are little things at my fingertips and I can't grab onto them. I have to reach into the distance because the shapes are smaller and further away from me."

On close inspection, this daily rhythm did not occur naturally but was caused by the pattern of Michael's ingestion of caffeine, nicotine, and alcohol, three substances that influence brain function.[14] His standard breakfast was cigarettes and volumes of coffee, both of which are well-known cortical stimulants. I conjectured that this morning stimulation dampened his synesthesia. In contrast, Michael drank heavily each evening, and his diary showed his synesthesia to be more vivid after several cocktails.

"You're describing a natural experiment," marveled Dr. Wood, momentarily pulling himself away from the dessert cart. "That's pretty neat. You could dry him out for a few days and then deliberately manage his exposure to stimulants and depressants."

"Yes," I smiled. "I'd predict that conventional stimulants would turn up the cortex and block synesthesia, while depressants should turn it down and make synesthesia more vivid."

When I next saw Michael he agreed to our carrying out this idea. We used one common medical stimulant, amphetamine, and two familiar cortical depressants, alcohol and amyl nitrite. The point was to see if external influences could alter the subjective experience of his synesthesia. Table 2 shows that such manipulations confirmed these predictions.

The stimulus for all our drug trials was an inhalation of spearmint oil. I already knew that spearmint produced a consistent synesthesia of smooth, cold, glass columns that Michael could palpate by running his hand along the back curvature and up and down the cylinder's length.

Table 2. Effect of various drugs on synesthesia		
Drug	Effect on Cortex	Effect on Synesthesia
Amphetamine	Stimulates	Blocks
Alcohol	Depresses	Enhances
Amyl nitrate	Depresses	Enhances

I quickly discovered that the drug amyl nitrite acted as an adjuvant. The term means "aiding," and describes any substance that boosts the effect of something else. We found that amyl nitrite greatly enhanced the synesthesia produced by spearmint. Not only was it more sensuous but the glass columns seemed to multiply so that Michael felt "hundreds of them. I can stick my hand in among the columns, touching and feeling their surfaces," he said. "They go back further than I can reach." Michael was of course eager to know why it did this, especially since amyl nitrite was then being used as a rec-reational drug called "poppers."

When my father practiced medicine, amyl nitrite was an old drug used for heart patients who had sudden chest pain called angina. By the late 1970s it had been thoroughly replaced by nitroglycerine tab-lets, but I always thought that amyl nitrite was medicine's most dis-tinctive drug. I say this not only because of the popping sound it made when you administered it (you did this by breaking the thin glass vial that was covered with cloth mesh and letting the patient inhale the vapor) but because of its unforgettable smell, something akin to aged sweat socks. There was never any doubt when my father used it on someone in his office, because the odor quickly invaded the whole house.

Michael was amazed that something external, like the poppers, could influence his synesthesia given that he was unable to alter it by his own will power. Despite his pleading to tell him how it worked, I had to keep silent until our work was finished. By itself, amyl nitrite did not cause synesthesia, but it was exactly what I had been looking for in a limbic-brain enhancer. Pharmacologically, it relaxes smooth muscle throughout the body, including the muscle in the blood vessels. As the vessels relax, their diameter increases and this causes the blood pressure to plummet. Although the heart increases its pumping in an effort to compensate, the net effect on the brain is a sharp but temporary drop in pressure at the end of the circulatory line, which happens to be the cortex. Without adequate blood flow, neural function abates. Amyl nitrite temporarily slashes blood flow to the cortex, suppressing it; this enhances the relative activity of the limbic brain.

As a recreational drug, poppers were most popular while disco dancing and during sex. The drug's clinical effects include a withdrawal into the self, slowing of time such that music seems slower or more distant, metamorphopsia, disinhibition, and heightening of emotion so that one appreciates the tribal aspect of dancing in an ecstatic throng. There is an oceanic state of oneness with the sexual partner, a sense of heightened orgasm, and an abandonment of judgment. Users routinely acknowledge performing sexual acts under the influence of amyl nitrite that they otherwise would hesitate to engage in.

The bottom line is that amyl nitrite causes one to dance "like crazy" at the disco and act "wild" in bed later on. "More is not enough" might well have been its users' motto. Its effects on behavior are consistent with a marked enhancement of sensual pleasures subserved by the limbic brain with a corresponding dissolution of higher judgment. Amyl nitrite has all the properties of a cortical solvent that acts within seconds and lasts only minutes. I use "solvent" to refer to a hierarchial dissolution of higher cortical functions such as social inhibition and logical reasoning. Alcohol has long been the best known cortical solvent that depresses inhibitions, which is exactly why it makes people loquacious at parties. With increasing amounts, however, individuals become more uninhibited as their judgment washes away.

▲■●

At a later time I gave Michael a dose of amphetamine and made sure the drug was exerting a physical effect by measuring his increased pulse and blood pressure. When he now sampled the spearmint, Michael was surprised to find his synesthesia very attenuated.

"It's like I'm grasping through a port hole," he exclaimed. "There is a small field of perception, with only one or two columns far off in the distance." He stretched his arm out. "It's very difficult to touch."

He stood there a moment, moving his hand back and forth slightly. "This is quite different," he concluded.

"How so?"

"The feeling comes faster than before but the columns are much smaller. They are still as vivid as a miniature would be compared to a large oil painting. The emotion is also less intense but still pleasurable. The whole thing is just more distant." He opened his eyes and looked at his fingers. "I can only explain it by saying that it feels like the sensation is slipping out of my hands."

"Let's try the amyl nitrite now." Michael was still stimulated by the amphetamine. I gave him a whiff of spearmint followed by the amyl nitrite adjuvant to see if it would counteract the suppressive effect of the amphetamine.

"This is incredible," said Michael. "The amyl nitrite doesn't work any more!"

"What do you mean it doesn't work?"

"This is not as visceral as it was when you gave it to me before. Instead of feeling it intensely in my hands, my back, neck, and arms, the feeling is centered on the fingertips. Before, I could focus my touch on a shape that was present all the time. Now, the feeling has become a series of small flashes of sensory things, only at the fingertips. The tactile quality is smaller, and I can't feel the quality of the column."

"You mean the coldness, the smoothness?" I asked.

"Yes. What have you done to me?" he demanded. Michael was visibly shaken. For all its strangeness, his synesthesia had been familiar. Now it felt alien, no longer a natural part of him. He sat at the table, trying to articulate his disbelief. "It's like a scale model, somehow, or a miniature. The whole thing is just incredibly different.

"The feeling is not sustained, but comes like pulses of a sensation, a sort of cinematic frame-by-frame sensation rather than being one long shape that I can concentrate on. The scale is still small and everything is far in the distance. The amphetamine keeps me outside of the columns and I can't get in to touch them, even with the amyl nitrite."

We tried other odors, flavors, even various foods while Michael was speeding from the amphetamine, yet none could induce his typical synesthesia. Even the amyl nitrite could not overcome the blocking effect of amphetamine despite many attempts to trigger synesthesia in this stimulated state.

Similar experiments with alcohol at a later time confirmed my prediction that it would increase the intensity of his synesthesia. A second, and unexpected, confirmation of alcohol's effect came several years later thanks to another "natural experiment." When I first met Michael in 1980 he was drinking about eight ounces of alcohol daily. This increased to nearly a fifth daily at the time he permanently stopped in 1985. When Michael's drinking stopped, so did his synesthesia.

Michael was horrified at losing it when he sobered up and he naturally wondered if his synesthesia hadn't been caused by his drinking all along. His loss was just an apparent one and there was a medical answer to the mystery, a phenomenon called rebound. This is a heightened nervous sensitivity following a removal of a brain suppressant from persons who have been chronically exposed to it. After alcohol withdrawal there is a rebound increase of cortical activity sometimes manifested by seizures, heightened autonomic nervous output (e.g., sweating, palpitations), insomnia, vivid dreams, nightmares, and tremors. In other words, withdrawal from alcohol, which is a cortical depressant, causes a rebound cortical stimulation, just as if one had been given a dose of amphetamine. Anyone who has experienced a severe hangover knows this jagged-up feeling. As time went by and the rebound effects of detoxification lessened, Michael's synesthesia returned, to his great relief.

Chapter 18

▲ ■ ●

JUNE 29, 1981:
BRIDE OF FRANKENSTEIN, REVISITED

"You want to inject radioactive gas into my head?" screamed Michael. "Are you out of your mind?"

"I haven't even told you what the benefits are."

Michael cocked his head back. "You can't be serious."

"For starters, they won't have to embalm you when you die."

Michael laughed and slowly shook his head. "Something tells me I'm already in over my head."

"Actually, I lied about the gas. The radioactivity saturates the tissues of your entire body, but I'm only going to charge you for the head." He grinned at me.

"Seriously," I resumed, "it really does go everywhere, although we are most interested in how the gas diffuses through your brain. By tracing the radioactivity we can measure how lively different brain regions are during synesthesia."

Michael frowned. "Who is 'we'?"

"We is yours truly and Dr. David Stump, a colleague of mine. He's a world expert in measuring brain metabolism. I studied the technique with him during my fellowship. Look," I continued, "you complained before that our experiments were too boring and not high-tech enough for your tastes. There are only about twenty of these devices in the world and we happen to have one right here. Dr. Stump agreed to let us use his equipment."

Michael shivered and screwed up his face. "It sounds so weird."

"This is your big chance for the fame you wanted," I cajoled him.

"I can't bestow the Tony Award, but I can promise when you're all wired up that you'll look better than the Bride of Frankenstein."

Michael rolled his eyes. "I'd still prefer the Tony. I think you had better explain it again," he said. "From the beginning."

I gathered my thoughts. "What I want is a probe to measure the level of metabolism in different areas of your brain while you have a synesthetic experience."

"Does such a thing exist?" asked Michael. "It sounds like the mad scientist. You *were* joking about 'Bride of Frankenstein,' weren't you?"

"Only partly. You'll be wearing a mask and a helmet that, quite frankly, does look right out of a B-movie. It has long probes and electric cables sticking out of it that carry a few hundred thousand volts across your head. You won't feel a thing, as they say, but to get back to your question, a way to measure localized brain metabolism using a tracer like radioactive xenon gas was conceived back in the 1950s. In the last ten years it has become a reliable tool of research neurologists."

"Why are we doing this?" asked Michael.

"For two simple reasons," I said. "First, I very much want to see if I can localize synesthesia. Neurologists have known for a long time that different brain areas contribute variously to different tasks."

"I know, localizing is your professional obsession."

"And yours is glamour, I understand. But let me continue. What made me think that the regional cerebral blood flow technique would be useful—let's call it CBF for short—was your experience with the amyl nitrite. Everyone knows that amyl nitrite reduces blood pressure, and we saw how dramatically it enhanced your synesthesia. Putting the two together, we should try to take a snapshot of your brain not only while you're having synesthesia but also when we boost it with the amyl."

"What kind of snapshot?" Michael asked.

"A living one. A CT or MRI scan only shows the structure of your brain; it can't say a thing about whether parts are working or not. But the CBF technique actually gives us a snapshot of the working brain, showing how lively its metabolism is in different places during a given mental or physical task."

"That's incredible," exclaimed Michael. "You can really do this in North Carolina?"

"No law against it," I said. In a quick shift to my sober mood I thought that maybe it still sounded too much like sci fi for Michael's comfort. "Maybe I had better lay it out step by step, starting with the idea of your brain doing work."

"You do that, while I fix a drink," Michael said, ambling over to the bar. "I have a feeling I might need one before we're done."

"Biologic work transforms energy," I started to explain. "Your skeletal muscle, for example, does mechanical work by lifting your weight against gravity when you climb the stairs. The kidney does chemical work when it osmotically filters the blood and concentrates wastes in the urine. The more biologic work done, then the more energy that tissue consumes. The brain, incidentally, is voracious. It consumes twenty-five percent of all energy used by the body."

"I just love an organ with a good appetite," Michael called from across the room.

"You really will want to do this when I'm finished," I answered. "If you want to know what parts of the brain participate in a particular task," I continued, "then you look to see which parts have the liveliest metabolism during that task. CBF is ideal because it can look at many brain regions simultaneously but independently. Whether you are squeezing a ball, reading, remembering, identifying shapes, calculating, or engaged in any other mental task, a few regions of your cortex will be metabolically more active than regions that do not participate in the task. Because the blood carries the necessary fuels of glucose and oxygen, it turns out that measuring the cerebral blood flow tells us how much metabolism occurs in any given spot of brain tissue. Areas working hard will have a higher blood flow than areas which remain at rest."

Michael became excited. "Does all this mean that you have localized where synesthesia takes place in my brain?" he asked. "Are you ever going to tell me what you found out?"

At this point it was permissible to divulge my hunches about synesthesia because Michael's knowing would no longer affect the results. I had finished with the experiments that required his being blind. I explained my idea of levels and how I expected that the cortex would not play a major role in the experience of synesthesia.

"This may sound like a stupid question," Michael said after I had filled him in, "but you said that the CBF technique looks at activity mainly in the cortex. But if you don't think the cortex has anything to do with synesthesia, then why are we using this machine?"

"Good question," I said, nodding slightly. "I agree that the logic doesn't appear to hang together, but the practical situation is that the machine we have is the machine we have. It's the most advanced high-technology probe that exists to examine *living brains*, and what it focuses on is the cortex. Secondly, I'm not expecting your cortex to shut down altogether when you have synesthesia," I told Michael. "I'm saying that the cortex won't be the hotbed of activity. What I do expect to see is a change in the pattern of overall energy use from the resting state, and that pattern should tell us something useful."

Michael stood silently for a moment. "So how are you going to convince me to go through with this horrible thing?" he asked.

▲ ■ ●

"Are you comfy?" I asked. Michael was strapped down on the couch, his head nestled inside the helmet with its electrical devices protruding two feet in all directions. Tons of nuclear, electronic, and computer equipment hummed around him in the darkened room. He looked like a cyborg. I was adjusting the strap on the black anesthesia mask that covered his face.

His eyes peered at me from beneath the helmet. "I'm not sure how I look," Michael mumbled through the mask, "but you were right. I *feel* like the Bride of Frankenstein."

The mask was how we attached Michael to the system. Through it, he inhaled xenon^{-133} gas, a naturally occurring radioactive isotope of the element xenon. It readily dissolves into the blood stream but, being biologically inert itself, causes no physical effects. Radiation detectors mounted in the helmet track the xenon-saturated blood flow from moment to moment in sixteen different brain regions while the subject engages in the task of interest—in Michael's case, his experience of synesthesia.

Blood flow measurements must be carefully planned because only so much radiation can be administered at a time. We chose a set of three sessions in one sitting. The first session would establish a *resting baseline*, against which we would compare two subsequent *activation* states. The baseline is measured in what is called the resting state: the patient lies on the couch, eyes closed, hearing only the steady white noise of the surrounding machinery. Obviously the brain is not "on hold" because random thoughts pass through the subject's mind, but this is nonetheless the standard procedure for the resting state worldwide. The normal pattern of resting blood flow should be fairly homogenous, with no brain region standing out from another.[15]

Measurement of Michael's resting state went well. After all, there was nothing for him to do except lie there. Dr. Stump and I were busy monitoring the huge volume of data that was collected by one computer in the room before it was sent on to the mainframe some distance away. Its mathematical analysis was complex and took much time; we would not be able to decipher the meaning of Michael's experience for about an hour, by which time we would have finished all three inhalation sessions.

I proceeded with the first activation task. The three of us did our parts quietly and deftly. I needed to keep Michael in a state of synesthesia for about eight minutes, the time needed to take my snapshot. A year-and-a-half's work had brought us to this point and I only had one chance to get it right. Michael and I had already decided on what stimuli we would use and how he would silently signal for more. The CBF apparatus and the mask best lent themselves to using a fragrance for the stimulus. We had previously determined that he could maintain a near-constant state of synesthesia for more than eight minutes without its wearing out. Speech, motor, and other sensory stimulation had to be avoided so that they would not cause the wrong areas to light up and confound the results.

Nearly any physical or mental task, even the administration of a drug, can be used as the activation state in the CBF test. Depending on its nature, you can sometimes intelligently guess which brain parts will "light up," or be more active, compared to the resting baseline pattern. Generally, you expect *at least a ten percent increase* in those regions that do participate in the activation task; twenty to fifty percent increases are common. Despite my long and deep thinking about synesthesia, I was chagrined not to have a really clear idea what to expect. I braced myself for surprises when we actually saw the results.

The third and last session, in which we boosted Michael's synesthesia with the amyl nitrite, also went without a hitch. The whole room was like a tranquility tank, dim and quietly humming, all superfluous activity forbidden. The deliberateness and sense of purpose with which we worked increased the tension. I allowed myself a moment to step back and observe. If what we were doing wasn't so earnest and fraught with risk, it would have been comical given how intense we were. I smiled at how oddly eerie and exciting it was, indeed just like those old movies. Once again, fact was as strange as fiction, perhaps more so.

"How do you feel? Was it good?" I hurried to disconnect Michael and extricate him from the apparatus, glad that we could finally talk after an hour's silence. Dr. Stump turned on the lights.

Michael sat up slowly. "It was so peaceful in there," he said, obviously surprised. "How strange."

"How was the synesthesia?"

"Oh, it was intense, don't worry. And it was practically constant, just like we rehearsed."

"Did the amyl nitrite work?"

"Like before. Except it was so extreme this time," Michael exclaimed, grabbing my arm. "Being in the dark, there was nothing else to catch my attention. I was so focused on what I was feeling. It was wonderful. Thank you so much."

The first part of our task had gone all right. Dr. Stump and I had constantly monitored the raw data and knew it was reliable. David fed a data tape cartridge into the computer console. The mainframe had crunched masses of numbers. Now it was time to decipher what they meant. "The first one is coming off line," he announced. "Come take a look at this."

Graphs and tables of numbers scrolled by on the computer screen while the printer hammered out hard copies. "Usually we wheel patients out on a stretcher," Dr. Stump joked with Michael. "They're not as young as you are. In fact, you're the first patient I can remember who has ever gotten to see this part."

"Which one is this?" asked Michael, pointing to the screen.

"Your baseline flow," answered Dr. Stump. "This is the state of your brain metabolism as you're resting on the couch, not doing anything. We'll compare the two sessions with synesthesia to this one here."

David Stump had been my teacher. He and I had looked through such data files many times before. We spoke jargon back and forth about initial slope indices, bicompartmental flows, standard deviations, and various clearance curves while Michael stood by patiently. But the message in our animated banter became clear to him.

"There's something wrong, isn't there?" Michael said at last.

"Yes," I said slowly. "I can't believe it. You're normal but it isn't."

Michael turned to look straight at me. "Could you be a little more specific?"

Dr. Stump tried to reassure him lest he misinterpret our debate. "It isn't abnormal in a bad way, Michael," he said, "but in an odd way.

What's remarkable about it is how variable your blood flow is from place to place. Normally the resting landscape is pretty homogenous. Also, your average flow for the whole brain is low for someone your age. In fact, the flow under some of the probes is so pathologically low that it approaches the limit of this machine to detect any blood circulation at all."

"To put it very bluntly," I interjected, "what David is saying is that some parts of your cortex have such minimal flows that they look dead. Obviously this cannot be because your neurological exam was perfectly normal. On the face of it, the conclusion is absurd. Perhaps the patterns during the activation states will reconcile this mystery."

"I hope so," said Michael. "For my sake," he added *sotto voce*.

Michael appeared so perfectly normal on the outside, neurologically speaking, and yet he was walking around with a bizarre metabolic landscape in his brain. As I wondered what it would look like during synesthesia, the answer scrolled onto the screen.

"This can't be right," David exclaimed after examining it.[16]

"I know, but it is," I shouted. "Look at the head and air curves. It's real!"

What we two scientists were hesitant to accept was that during synesthesia, the average blood flow in Michael's left hemisphere dropped to three times below the lowest acceptable limit of an average person's! In some places he had absolute values of flow that Dr. Stump and I had never seen in anyone who did not have obvious neurological symptoms at the time of their study. The flow in Michael's left hemisphere was *eighteen percent less than it was in the baseline.*[17] And yet other than having synesthesia, he had no subjective or objective manifestations at all.

"What kinds of symptoms should I have?" asked Michael after I told him this.

"Well, you should be blind, or paralyzed, the usual kinds of neurological symptoms. It's hard to appreciate what this means," I emphasized, "how exceptional your study is."

"Why? he asked.

"An overall decrease *never* happens during the activation task of a typical study," I explained. "If we pulled someone in here off the street, it would be impossible for us to reproduce the reduction of blood flow—and therefore the reduction in the cortex's metabolism—that you have during synesthesia."

"He's right about that," Dr. Stump chimed in. "Just about everything *increases* flow. I only know of two drugs that could reduce it at all, and then they might do so by ten percent at most. Personally, there are a lot of interesting technical details about your study, Michael, but the most important point is that instead of an *increase* in metabolism, which is what I expect with any kind of activation, your brain shows a *profound decrease* in cortical metabolism during synesthesia. All the change is taking place in the left hemisphere, too. This was also true when we gave you the amyl nitrite adjuvant.

"Frankly," he said, "I've been doing this for fifteen years and I've never seen anybody remotely like you. I'm astounded that a smell, a mundane little sensation, can cause such a massive redistribution of metabolism in your brain, one that essentially turns off your cortex." He shrugged. "Why you have no neurological symptoms baffles me."

"I agree. These results are stunning. There's no question your brain is fascinatingly different," I told Michael. "If you wanted objective proof that synesthesia is real, here it is. It doesn't get much more real than this."

"But what does it all mean?"

"It means without a doubt that synesthesia does not occur in the cortex," I said. "My theoretical ideas were correct. But the magnitude is astounding. The blood flow data are incontrovertible. When you experience synesthesia, your cortex shuts down in a way I never thought possible. Researchers all over the world have studied thousands of individuals with this technique, both ill people and healthy controls. Your brain is dramatically different from anyone we know who has been studied with this technique."

"So you failed," Michael said quietly. "This shutdown of the cortex means that it's impossible to localize synesthesia, is that it?"

"Good grief! Not at all," I said. "We've localized it all right. All the interesting change takes place in the left hemisphere, but not in the cortex."

"The radiation counts show that there is an overall increase in brain metabolism as a whole," Dr. Stump added. "There is definitely metabolic work going on when you feel the synesthetic shapes."

"I'm confused," Michael said. "If overall brain metabolism increases but the cortex shuts down," Michael asked, "then where does the energy go? Where is synesthesia taking place?"

"In the limbic brain," I smiled.

▲ ■ ●

Michael's study showed how science is often counter-intuitive, since the standard view of brain function has ingrained in us the assumption that every mental state is represented in the cortex. Popularizers too have contributed to its glorification. One forgets that the majority of the brain is *not* cortex at all, but other types of tissue. You start to assume that the remainder of the brain acts only to prop the cortex up on the surface. An analogy is expecting a gorgeous wedding cake to be no cake at all, but only fancy frosting held up by cardboard and tinfoil.

The human brain is not just cortex held up by inert support. To pursue the metaphor, the gray matter "icing," which is the visible part on the surface, is only 1 to 2 mm thick, a fraction of the brain's total volume. Unlike my metaphor, however, there is plenty of cake below. Subcortical tissue is not just there to hold up the surface; it does a huge amount of biochemical work, most of it outside of consciousness. I already referred to the gustofacial reflex and how brainless (anencephalic) infants are nonetheless able to discriminate sensations. Comparative neurology shows sophisticated behavior in animals that have no cortex to speak of (such as birds). Suction removal of fairly large amounts of monkey cortex results in animals difficult to distinguish from their cage mates. This is one of many lines of evidence suggesting that the cortex provides a fine grain of discrimination to those brain entities that ultimately decide behavior.

The final arbiter is none other than the limbic system, buried deep within the temporal lobe. It is deep enough that its metabolic activity is beyond the range of the blood flow method to measure it. Yet given the stunning shut-down of the cortex, the many pieces of evidence I had gathered over the years pointed here as the seat of synesthesia.

Chapter 19

▲ ■ ●

HOW THE BRAIN WORKS: THE NEW VIEW

Renaissance astronomers kept piling epicycle upon epicycle to explain the retrogression of Mars's orbit until their scheme of planetary reaches was a patchwork that no longer held together. Kepler's conceptual shift of planetary orbits from circular to elliptical explained all observed facts much better, and without the need for piling on special exceptions.

The old view of brain organization has collapsed in a similar fashion: it could no longer bear the weight of having to explain the avalanche of observations made during the recent past. The standard view of how the brain works, which I outlined in Chapter 4, is a product of the nineteenth century. We now know that as a model its generalities are true, while some of its specific predictions are erroneous.

Neuroscientists have just lately come to realize how important emotion is. Placing reason and the cortex first and foremost really overstates the case. Like the Wizard of Oz shouting, "Pay no attention to the man behind the curtain," reason and an accomplice called self-awareness have deluded us into believing that they have been pulling the strings. But we shall soon see that emotion and mentation not normally accessible to self-awareness have been in charge all along. Some wonderful discoveries are in store.

But first, a quick reminder of the standard view and the triune brain.

The Standard View, Revisited

To recapitulate the three points of the old view: (1) the flow of neural impulses is linear and hierarchic, (2) physical and mental functions are localizable to discrete regions of the cortex, and (3) their hierarchical arrangement implies that the cortex dominates everything else. A conclusion drawn from these three principles is that the brain's cortex is the seat of the human mind. Cases such as Phineas Gage's seemed to prove that here is where consciousness and reason are located.

According to the old view, synesthesia should have been located in the cortex, most likely the parietal lobe's tertiary association area where the three senses of vision, touch, and hearing converge. (Remember that taste and smell were dismissed as unimportant.) We approached synesthesia historically, descriptively, and experimentally: the cumulative results made us suspect that the above explanation was wrong, and the cerebral blood flow measurements blasted it away for good. How synesthesia really worked turned out to be counterintuitive: all the obvious possibilities were wrong because the old view of how the brain works was wrong.

How sensory flow actually proceeds from the outside world to our mental one inside is vastly different from popular accounts of "how the brain works" that are rooted in the old view. The old view emphasized the cortex as the highest entity. As I will show momentarily, the contemporary view puts the role of the cortex not at the top, but more in the middle of multiplex, parallel, and recursive pathways. The word *highest*, when applied to the word *cortex*, is a meaningless attribute. Cortex is just one of several types of brain tissue. Intense but recent interest in consciousness and emotion, after decades of disinterest, has led neuroscientists to conclude that the limbic system, not the cortex, exerts the greater influence.

The Triune Brain, Revisited

Paul MacLean's triune brain, shown in Figure 2, page 21, has captured popular attention during the last forty years for several reasons, one being that it is easy to understand. Today we know that parts of his idea are wrong, and that it is best taken as a useful metaphor rather than as an exact model of brain organization.

The triune brain did help to show that specific categories of behavior could be assigned to different types of brain tissue, each of which had a unique evolutionary history. It was enormously helpful in showing that tissue *below the cortex* was not just inert filler that could be neglected, but that subcortical tissue was enormously important to behaviors that could not be dismissed as merely "instinctive" (reproduction, feeding, and fight-or-flight situations, for example). In general terms, the behaviors in question include grooming, routines, rituals, hoarding, protection of territory, deception, courtship, submission, aggression, socialization, imitation, and many other human proclivities.[18]

MacLean coined the term "limbic system" in 1952 because of the extensive relation of entities below the cortex to the limbic lobe of the brain. Broca first defined the limbic lobe in 1878 in structural terms, as the inside rim (*limbus*) of the hemispheres where they meet the brainstem. MacLean's extensive experimental work showed that one part of the limbic system was concerned with preservation of the species (being the substrate for behaviors related to sex, procreation, and socialization), while another part was involved with self-preservation (feeding, fear, and fighting). He identified a third segment as controlling suckling, maternal and paternal care, audio-vocal communication, and play—behaviors that are distinctive of mammals.

Unfortunately, MacLean's conceptual drawing of the triune brain had an unintended effect. Because he pictured the neomammalian brain (the cortex) enfolding everything else, people still ceded it too much importance. The elevated role in human behavior that MacLean assigned to subcortical tissue (as represented by the reptilian and paleomammalian brains) was unintentionally undermined by his drawing. People came away with the hierarchical notion that the cortex was still the boss.

Whether a neural structure is visible on the surface or tucked out of sight has no bearing on hierarchy or whether it is controlling or controlled. Function is what matters. As the new view will clarify, the complexity of anatomic connections between cortex and subcortical entities are reciprocal and thus interdependent.

THE NEW VIEW OF HOW THE BRAIN WORKS

The new view has five main points, which I will develop. In summary, they are as follows: (1) The flow of neural impulses is not linear, but parallel and multiplex, including transfer of information that does not even travel along nerves. Thus, it makes no sense to speak of hierarchy. (2) We no longer speak of localization as a one-to-one mapping, but of the distributed system: a many-to-one mapping in which a given chunk of brain tissue subserves many functions and also, conversely, by which a given function is not strictly localized but is distributed over more than one spot. (3) While the cortex contains our model of reality and analyzes what exists outside ourselves, it is the limbic brain that determines the salience of that information. (4) Because of this, it is an emotional evaluation, not a reasoned one, that ultimately informs our behavior. (5) Likewise, all analogies of the mind to a machine are inadequate because *it is emotion, much more than reason, that makes us human.*

Non-Linear Information Flow and "Inner Knowledge"

Consider this assertion: you know more than what you learn through rational thought and language, although you most likely are unaware that you know it. This reference to what is usually called "inner knowledge" is conventionally made by humanists and spiritually sensitive persons. Those who believe themselves to be objective may be uncomfortable accepting this premise without scientific proof. Human neurology gives credence to this assertion because information is actually transferred through the nervous system in more ways than most people realize. Multiple communication channels exist in addition to what we typically know of as nerves, synapses, and the circuitry familiar from classical neuroanatomy. This abundance of alternate routes is denoted by the word "multiplex."

The multiplex ways of transmitting information in the brain is not hierarchical, as would be if the flow were straightforwardly linear, but involves connections which are parallel, recursive, feedforward, and feedback. There exists a wide assortment of molecules, such as hormones and peptides, that also act as information messengers. More than fifty are known and more are being discovered every year, not just in the brain but throughout the body. Information can therefore be transmitted throughout the entire body not only by neurons and axons (the traditional long wiring system of the brain), but through the extracellular fluid that surrounds the whole system itself. This method of communication, called *volume transmission*, is already the topic of several books.[19]

Think of electrical transmission along nerves as a train traveling down a track; volume transmission is the train leaving the track. The idea that information is communicated by molecular messengers at rates that can be very fast (up to 120 meters per second in axon fluid) or slow (e.g., diffusion of peptides in cerebrospinal fluid) has opened up our understanding that the human brain has systems that are much more complex than we had ever supposed, communicating at different ranges, different velocities, and different methods.

One of the reasons that current approaches to artificial intelligence (AI) have failed is that they try to imitate logic and are modeled on the circuitry found in the cortex. They do not take into account that the biologic brain has many different ways of transferring information. To be successful, AI would have to incorporate some kind of regulating system to organize all the different means of information transfer.

In the human brain it is the limbic system that performs this regulation, a fact only recently confirmed in 1985. You might wonder why it has taken until now to figure out something that seems so fundamental. The reason is that only recently have anatomic techniques been developed that permit neurotransmitter molecules to be tagged with special dyes that can be seen at the microscopic level. One can now follow the journey of neurotransmitters both through nerve fibers and the extracellular spaces in which volume transmission occurs. Classical anatomy allowed us to map the brain's general circuits, but accurately knowing the *direction* of flow for the first time as well as the precise origins and targets of various transmitters has forced revisions. It turns out that every single division of the nervous system, from the frontal lobes to the spinal cord, contains some component of the limbic system. In other words, the limbic system forms an *emotional core* of the human nervous system.[20]

Function Is Not Strictly Localized

The idea that circuits rather than "control centers" support the expression of emotion was first suggested by James Papez in 1937.

The major entities of what we now call the limbic system were hooked together into the Papez circuit, through which all aspects of emotion were manifest. The implication for the neurologist's habit of localization was profound: emotion was no longer localized in a discrete control center but was spread out over pathways. Of course, the pathways must be somewhere, and so some localization is involved. But it is qualitatively a much more diffuse localization than that imagined by the old view.

By the late 1970s, this approach had fundamentally and permanently changed how we conceived of information traveling through the brain, although this new view has still not been disseminated to a large audience. The linear idea of discrete work stations along a conveyor belt gave way to the concept of multiple mapping in which a brain with multiplex communication channels can process information in many locations at once.

Multiple mapping is possible by linking one input to several outputs. As soon as nervous impulses from the sense organs synapse in their respective primary sensory cortices, they simultaneously branch out to *multiple* areas of association cortex for further computation, each area being concerned with a *different* facet of the experience. In vision, for example, each of some *two dozen* areas handles a different aspect of seeing. The job of analyzing whatever it takes to yield the experience of color goes in one direction. The many things that constitute shape and lead to the recognition of an object, or the space where that object is located, are handled somewhere else. The neat image impinging on our retinas is shattered as the world is multiply mapped in our brains, a different map in each of several areas per sense. In addition, collateral branches shoot out to form recursive feedback and feedforward circuits that contribute to a massively parallel computation.

The ability of a discrete brain area to process several *uniquely different maps* of the world arises from the complicated pattern of its inputs and internal connections in each architectural region, and the linking of this calculation to *several* outputs. This is what we call the distributed system, which means that the many aspects of complex functions (e.g., vision, hearing, memory, or emotion) subserved by a particular circuit are not rigidly located in any one of its segments, but rather in the dominant process occurring at any given time in the circuit itself. The number of distinct regions transmitting to and receiving from other cortical areas varies from ten to thirty. The exponential level of complexity is apparent, and far beyond that inherent in the conveyor-belt progression of the old view.

The Primacy of Emotion

The complexity of multiple mapping is one reason why we conceive that the role of the cortex is to analyze the external world and contain our model of reality. It is the limbic brain, however, that decides questions of salience and relevance and so determines how we act on the information we have. It is an emotional calculation, not a logical one, that animates us. The development of mammalian brains shows us why.

For centuries, emotion has been looked down on as primitive, and reason held as the superior development. Let us look at some possible physical reasons for this belief.

Environmental and cultural changes do not directly affect the direction of evolution. In other words, acquired traits such as knowing how to speak German or having a broken nose are not passed through genes to your offspring. What environmental changes do create is different niches for adaptation, and these niches are filled up by the natural selection of organisms with favorable genetic mutations. Before the development of big-brained human beings, change occurred through the force of evolution acting slowly over eons. Once brains develop any kind of organizing system for memory, with its ability to influence present and future actions, we immediately leap beyond the slow gene-by-gene mutation that physical evolution allows. This is because cultural change accumulates rapidly and can be passed on to others, not genetically, but through cultural transmissions (which, for example, is happening as you read this book).

This is an oft-cited advantage of having big, well-developed brains as well as being one explanation for why humans have achieved so much in terms of culture and technology compared to other vertebrates. Our large frontal lobes are especially given credit for this. Since the cortex appears to have developed far out of proportion to that tissue below the cortex, we usually point to the cortex when we say, "This is what distinguishes us from less advanced species." The flip side of this assumption, however, is precisely why emotions are regarded as primitive.

The common assumption that the developed cortex makes humans unique implies that human limbic structures are no different from those of other mammals. If true, then human emotions are comparatively primitive. But exhaustive anatomic studies such as those I mentioned above show that the limbic system was not left behind by evolution. Limbic and cortical circuits co-evolved, and so reason and emotion burgeoned together in tandem.

Vertebrate brains in general started to develop in the reptilian line. But with the appearance of early mammals, the limbic system underwent major changes. It is now a constant feature in all mammals and is not seen in its *developed* form in submammalian species. Its robustness reaches an apex in humans, *making the human emotional system more powerful than that of other mammals. Additionally, the processing of emotional information is qualitively different from the processing of other information.* Is it then not likely that emotion therefore plays a larger role in our lives than we have hitherto suspected?

There is no doubt that the cortex and limbic system are reciprocally connected, and that one system can influence the other. The real question is: Does the influence of one outweigh the other? The fate of the spiny anteater from Australia can shed some light on this important question.

The Australian anteater, *Echidna,* is a prosimian low down on the mammalian family tree. Its tremendously developed frontal cortex is far larger than that of primates, our much closer relatives. If we had frontal lobes as massive as the spiny anteater's per unit volume of brain we would have to carry our heads in front of us in an extra-large wheelbarrow. This contradiction of a simple animal with a proportionately huge frontal cortex suggests an evolutionary strategy that went down a dead end. Having more analytic space for more computations does not make an efficient brain nor one that is especially smart.

The Australian anteater also appears not to dream, illustrating the general principle that for every gain there is usually a loss elsewhere. It is as if the anteater paid for its massive frontal lobes of cortex by forfeiting limbic function. The evidence is this. All other mammals dream, and when we do we emit an electroencephalographic signal in REM sleep called the theta rhythm, which is particularly prominent in the hippocampus of the limbic system. The anteater has no theta rhythm and presumably does not dream. When any sensory signal comes into the human cortex, the hippocampal theta rhythm becomes active if the stimulus is evaluated as relevant and you attend to it. Earlier, I used the word "salience" to describe a prime limbic function. The word means to "leap up" or "stick out," which is exactly what relevant stimuli do to grab our attention. That the hippocampus and the limbic system are a gateway, or valve, to perception is a crucial point I shall expand in the *Part 2* essay, "Consciousness Is a Type of Emotion."

By looking at the direction of flow and the scope of connections with their new methods, authorities in neuroanatomy have clarified that the hippocampus is a point where everything converges. All sensory inputs, external as well as those from our visceral, internal milieu, must pass through the emotional limbic brain before being redistributed to the cortex for analysis, after which they are returned to the limbic system for a determination whether the highly processed, multi-sensory information is salient or not. If so, we will likely act; if not, we will ignore it just as we do the majority of the irrelevant energy flux that constantly bombards our nervous system. In determining salience, the emotional brain acts something like a valve, deciding what will grab our attention and what will not.

It turns out that the brain's largest and latest development, the cortex, has more inputs from the limbic system than the limbic system has coming from the cortex. *The functional significance of these connections turned out to be the reverse of what we had assumed for decades.* Granted there is a reciprocal relationship between the cortex and the limbic system, each regulating the other, and each ultimately influencing our mental life. But the number and nature of the recursive feedback circuits ensures that the influence of the limbic system is greater.

Let me close with two clinical examples of how emotion rather than reason is primary. The first concerns patients in coma. In the sequence of recovery they first manifest automatisms, then voluntary movements and speech that is childlike and emotionally childish. If recovery continues, their behavior becomes what we would describe as more rational and adultlike. The pattern of recovery from coma shows that intellect cannot be reclaimed unless emotion recovers first.

In discussing temporal lobe seizures that originate in the limbic system, I mentioned that they can produce involuntary actions (automatisms) that seem purposeful but for which the person has no awareness or recollection. TLE can also cause compulsive thinking, florid psychosis, and episodes in which one cannot distinguish between dreaming and reality. The overlap between the behavior of TLE and that of psychiatric disorders is striking: fifty percent of those with temporal lobe seizures show psychiatric symptoms compared to only ten percent in other types of epilepsy. Thus the emotional brain is physiologically able to overwhelm the rationality of the cortex.

We can conclude that emotions play an important role in our behavior, perhaps a role even greater than that of reason.

Chapter 20

▲ ■ ●

THE IMPLICATIONS
OF SYNESTHESIA

Pursuing the puzzle of synesthesia led us to realize that reason and emotion co-evolved. This position weakens the strong dichotomy commonly perceived between reason and emotion. It suggests that emotion has a "logic of its own" and must be understood on its own terms.

When Dr. Stump and I stood with Michael Watson in the cerebral blood flow lab, we concluded that synesthesia is localized to the limbic system of the left hemisphere. In itself this is an intriguing deduction, but one that I feel remains incomplete. Let us now be more specific about two things: (1) how synesthesia works, and (2) what its implications are.

HOW SYNESTHESIA WORKS

Putting it as plainly as possible, parts of the brain get disconnected from one another (as they do in release hallucinations), causing normal processes of the limbic system to be released, bared to consciousness, and experienced as synesthesia. In other words, a stimulus causes a rebalancing of regional metabolism. This explanation is analogous to the very common condition of migraine, but before clarifying the analogy let us examine each part of the claim.

First, sensory stimuli can and do cause large changes in focal cerebral metabolism. This much is known from long experience with techniques that measure function, such as the EEG, cerebral blood flow, or positron scanning. These observations simply restate the axiom that some parts of the brain are more active than others depending on its physical or mental task. Secondly, changes in focal metabolism necessarily alter the relative strengths of connections among the various entities of the brain, some being strengthened while others are reduced.

These first two premises of how synesthesia works are consistent with standard understanding of brain physiology. In release hallucinations we saw how the function of one entity could be released from the influence on another by destroying it; changing the strength of connections between them can produce the same result. In this case, the release will be temporary, as synesthesia is.

This premise was vividly demonstrated during Michael Watson's CBF. His highly unusual pattern of blood flow settled the issue that synesthesia is not a cortical function. However, it also raised an alarm that something was seriously wrong with his brain, and I considered it medically necessary to see whether he had a malformation, tumor, or other problem with the blood vessels in his brain by examining them with an X-ray called an angiogram.

His angiogram was completely normal, a finding that deepened the mystery of synesthesia. However, a valuable clue emerged during the procedure. In order to make blood vessels visible to X-rays, inert contrast fluid is injected into them. Blood normally brings oxygen and glucose to the brain tissue. Therefore, replacing blood with contrast fluid means that the tissue is without either fuel for about ten seconds. Patients usually feel a flush of heat when the contrast fills their arteries, but otherwise have no symptoms. That is, their brains are well able to tolerate this momentary lack of fuel. Michael's brain, however, could not tolerate it. I was flabbergasted when he had visual, auditory, and tactile perceptions as the contrast fluid was injected. This clue suggested that the energy metabolism of his cortex is precariously balanced and that he has no reserve when taste and smell stimuli drastically alter his cerebral circulation, causing an effective disconnection and a release of synesthesia to consciousness.[21]

Consistent with the results of the cerebral blood flow examination, perceptions occurred only when the *left* side of Michael's brain was injected. He saw, for example, "an intense pink and the blackest black I have ever seen, flashing like lightning." A ringing in his left ear was "a high-pitched whine, higher than a siren." He felt an excruciating "bone pain, like a toothache" in his neck and the back of his head. The episode ended with an image of overlapping geometric cubes, like Art Deco, that rapidly alternated between black and white, multiplying and "growing like crystals." These things never happen to the several hundred thousand patients who have cerebral angiograms each year.

I suggested that the explanation of synesthesia was similar to that of migraine. Unfortunately, the word "migraine" has become corrupted in ordinary language to incorrectly refer to any severe headache. Neurologists precisely refer to the syndrome of classical migraine as an aura of sensory distortion quickly followed by an excruciating headache on one side of the head. The sensory aura is most often visual, and consists of photisms similar to those of synesthesia or form constants. Tactile, auditory, gustatory, and visceral auras are less common.

Textbooks explain migraine's sensory manifestations as caused by a spreading depression, like a ripple in a pond, of both the brain's blood flow and its electrical field (the two are coupled). This explanation has been standard since the mid-1800s, although no cogent answer exists as to why such a thing occurs in susceptible persons. Since this explanation is repeated in countless textbooks, no one ever objects to its failure to explain migraine better. The explanation is appealing, simple, and easily grasped by students and patients.

I am not saying that the physiology of migraine and synesthesia are the same, but analogous. I compare them to answer those who might object that my explanation of synesthesia is not sufficient. The only answer I can give is that it is as good as our textbook explanation of migraine is. The analogy between them is as follows. Synesthesia is evoked by a stimulus, and migraine likewise has an identifiable trigger (e.g., foods, scents) in some persons. We accept as fact that metabolic and circulatory alterations do occur in migraine for no apparent reason and then resolve, also for no apparent reason. Like synesthesia, migraine can also be influenced by drugs. Even though it is one of the most frequent conditions in neurology and its cerebral physiology is well understood, we do not understand why migraine exists any more than we know why synesthesia exists.

So, asking "Why do only some people experience synesthesia?" is like asking "Why do only some people have migraine headaches?" or any other condition. I suggest that the proper question is "Why are some people consciously aware of synesthesia?" I put it this way because after studying this marvelous phenomenon for over a decade, I have come to the opinion that synesthesia is a very fundamental mammalian attribute. *I believe that synesthesia is actually a normal brain function in every one of us, but that its workings reach conscious awareness in only a handful.* This has nothing to do with the intensity or degree of synesthesia in some people. Rather, it is that most brain processes operate at a level below consciousness. In synesthesia, a brain process that is normally unconscious becomes bared to consciousness so that synesthetes know they are synesthetic while the rest of us do not.

▲ ■ ●

Bearing in mind my comments about the brain's multiplex communication and its distributed systems, I have identified the hippocampus as the main node that makes synesthesia possible. The hippocampus is a major component of the limbic system and resides in the temporal lobe, tucked away next to the brainstem, literally folded beneath the cortex. The limbic system itself is very old and connected to just about everything. Some indication of its extent and importance can be gleaned from my own favorite brain atlas, which takes twenty-four separate illustrations to convey its pertinent physical features and forty-three pages of text to summarize its connections.[22]

Our review of the six altered states of consciousness first suggested that the hippocampus is involved in the perception of subjective experiences. A more direct reason why it is essential to the experience of synesthesia comes from observation of persons who have a seizure originating in the hippocampus and who have synesthetic experiences relatable to that seizure but are otherwise not synesthetic.[23]

The strongest reason for singling out the hippocampus is anatomic. Only here is it possible to bring together information that was processed in functionally and geographically separate parts of the brain. These gathered signals flow into a unique structure that also knows about the internal milieu as well as the fundamental drives of the organism as a biological entity. The hippocampus can also respond back to virtually every entity that originally fed into it, including autonomic structures that govern the internal milieu. It is here that the autonomic responses could add the pleasure that synesthetes feel during their multisensory experiences.

THE IMPLICATIONS OF SYNESTHESIA

Synesthesia is a conscious peek at a neural process that happens all the time in everyone. What converges in the limbic system, especially the hippocampus, is the highly processed information from sensory receptors about the world, a *multisensory evaluation of it.*

I call synesthetes *cognitive fossils* because they are fortunate to retain some awareness, however slight, of something that is so fundamental to what it means not only to be human, but mammalian!

Could we possibly evolve into a class of synesthetes who have this added human capacity? The answer is that we already have, except we don't know it. Synesthesia is not something that has been added, but has always existed. A multisensory awareness is something that has been *lost* from conscious awareness in the majority of people, which again is what forces me to think of synesthetes as cognitive fossils.

We know more than we think we know. The multisensory, synesthetic view of reality is only one thing that we are sure has been lost from consciousness. There could be a lot more. If you want to try to reclaim some of this deeper knowledge, I suggest that you start with emotion, which to me seems to reside at the interface between that part of our self which is accessible to awareness and that part which is not.

Mammalian brains show two independent evolutionary trends in the expansion of surface cortex and subcortical limbic tissue, with most species high in one or the other. Monkeys, for example, have substantial cortical development but little limbic enlargement; rabbits show the opposite trend. Humans are unique in being advanced in both limbic and cortical dimensions. I have already argued that limbic pathways did not get left behind in evolution as the cortex expanded. The two co-evolved. In fact, our number of limbic fiber tracts is greater both in relative size and absolute number compared to other fiber systems. For example, humans have five times as many fibers in a single limbic pathway called the fornix than in both optic tracts that bring in visual information from the outside world. It is the limbic system more than the cortex that reaches its greatest development in humans. It also happens to be most closely associated with those emotional and subjective traits we identify with humanity.

The basic pattern of the limbic system has remained constant throughout its evolutionary embellishments. In all living vertebrates one finds considerable uniformity in its components and their connecting circuits. Despite evolutionary changes in itself and other brain components, the limbic brain remains the terminal stage of information processing, that stage for suppressing automatic, habitual responses in favor of new alternatives when the unexpected happens. The limbic system gives salience to events so that we either ignore them as mundane and unimportant, or take notice and act. It is also the place where value, purpose, and desire are evaluated, a process referred to as assigning negative or positive "valence."

The function of calculating valence could have had one of two evolutionary fates. It could have been assumed by the cortex so that questions of meaning and purpose are evaluated by a more analytic and presumably detached organ (what people call "objective"). The other way the function for valence could have evolved is the way it did, which has been often misunderstood. The limbic brain has retained its function as the decider of valence. What the cortex does is provide more detailed analysis about what is going on in the world so that the limbic brain can decide what is important and what to do. The choices boil down to fundamental ones about what it means to be a living organism.

I am not suggesting that individuals who make their choices "emotionally" are more human than those who claim to be rational. Since we happen to have the best integration between the motivating force of the limbic brain and the analytical one of the cortex, it makes more sense to say that persons who balance reason and emotion are the most human, since they are using *both* systems which define the human neurology to the fullest.

For most people, I suspect that the best advice is to permit the intellect to only inform your choices, not to override your fundamentally emotional ones. My claim is that we have grossly neglected the importance of emotion in our lives. Through reason, you may deduce that there is a logic of emotion and accept the conclusion that it is the major force that guides your thinking and action.

The Illusive Nature of Self-Awareness

For those who demand "objective" proof of the illusive nature of analytic awareness and my assertion that something other than that entity we call our "self" is in charge of our minds, I point to the work of Kornhüber.

It may be difficult to bring yourself to believe that you are more emotional than logical, let alone accept the assertion that the entity you know of as your self is not really in charge of your mind and the direction of your life. I can only point to the large literature in philosophy, neuroscience, and cognitive science that challenges this belief. The distinction between reason and emotion seems at first to be based on self-awareness. But there is more to the distinction than a binary choice between self-awareness or the lack of it (unconsciousness). A physical parallelism may be useful here.

The physical brain has no sensory nerves and is not aware of its own physical substance. One can mechanically poke it with a rod, or stimulate it with an electrical current or strong magnetic field. Regardless of the type of physical energy applied, the patient does not say, "I feel you touching my brain now," but rather reports a sensation in a peripheral body part. Physical movements may also be elicited by stimulating brain motor parts. This immediately tells us that *there is not an identity between the actual neurons being stimulated and the spatial location of the subjective experience that results from it.* Repeated experience from brain surgery on awake individuals confirms that the brain's reference is external.

People who have their brains stimulated have experiences that they claim "they" did not cause. They know rationally that the experience comes from them even though it does not feel like it comes from them. It is like getting a view from somewhere and a view from nowhere combined. Instead of being aware of its own physical substance, the brain is aware of a reality it creates outside of itself. This world of the not-self appears to be an independent world of objects, space, and time. A helpful analogy may be to think of a hologram, which produces a real three-dimensional image that exists outside of the film and laser beam that produce it.

The illusive nature of self-awareness was first shown by Kornhüber in 1965.[24] His work on what is called the "readiness potential" shows that there is more to consciousness than what is apparent from either what we can introspect or observe. The experimental subject is told to move a finger whenever he or she feels like it. Devices measure precisely when finger movement occurs as well as electrical potentials in the brain before and after the movement. A special clock permits the subject to record the moment of conscious decision to move the finger. Through a method called reverse analysis one finds a buildup of brain activity, the readiness potential, that prepares the action to be performed. This occurs in the brain almost one second before the subject makes a conscious decision to move the finger. One second is long on the time scale of physiologic events, much longer than the time required for the electrical impulse to traverse the physical nerves between motor output and the finger muscles. In other words, the readiness potential far antedates the subject's *decision* to make a movement.

Others have replicated and extended Kornhüber's work.[25] The conclusion is that we are deluded in believing that each of us is a free agent who may decide to take an action. Such a decision is an interpretation we give to a behavior that has been initiated some-place else by another part of ourselves *well before* we are aware of making a decision at all. In other words, the decision has been made before we are aware of the idea to even make a decision. If "we" are not pulling the strings, then who or what is? The answer is, it is an unknown part that is unfathomable to introspection.

Kornhüber's experiment can also be performed in the other direction. A stimulation of the skin with a pin prick reaches the brain in ten thousandths ($^{10}/_{1000}$) of a second; but there is more than half a second of readiness potential before the pin prick is consciously perceived. Yet in the subject's own mind there is paradoxically no delay in appreciating the prick relative to the physical stimulation. Benjamin Libet, a contemporary physiologist exploring consciousness, proposes that the initial signal in the cortex, occurring just ten thousandths of a second after the pin prick, acts as a time marker. After the half second ($^{500}/_{1000}$) of neuronal buildup necessary to convey a conscious experience, that experience is referred back to the time marker. The subject thus experiences no delay in his or her own mind. Personal experience tells us that the world appears seamless, that the sights, sounds, smells, tastes, and textures of any experience are concurrent. Libet suggests that the unsettling time disparity between conscious experience and neural events is necessary to maintain subjective synchrony among sensations. The point is that subjective experience has a discrepant time base from the neuronal activity that produces it.[26]

Our conscious self is the tip of an iceberg. Work such as that done by Kornhüber and Libet gives testimony to the existence of a part of us which is inaccessible to the self. The implication of their work is consistent with that of split-brain research. For example, the "conscious," speaking hemisphere of split-brain patients is surprised at the knowledge and actions of the other hemisphere. This is particularly striking when interhemispheric conflict occurs, as when the left hand undoes a task that the right hand has just completed, or when the left foot goes in one direction while the rest of the body tries to go in another.

These kinds of mind probes show that we do know more than we think we know. And yet isn't it interesting how we are always surprised to discover that we know it? In everyday life are we not surprised at our own intuition, creativity, artistic inspiration, insights, and other manifestations of inner knowledge? Unfortunately, people rarely trust this inner knowledge because we are told in so many ways to trust in objective facts instead.

I conjectured that synesthesia was a normal brain process, but one that was hidden from the consciousness of all but a handful of the 5.4 billion humans on this planet. I also suggest that the altered states of consciousness I discussed may actually be moments when the "real" us comes to the surface. Things that "we" do not do but which instead "happen" to us, things such as emotions, insights, intuitions, or feelings of certitude, are created by a facet deeper than the one with which we are cognizant.

What we think of as voluntary behavior, set in motion by free will, is really instigated by another part of ourself. Part(s) of us are *inaccessible to self-awareness*, the latter being only the tip of the iceberg of who and what we really are. The "I" is a superficial self-awareness constructed by our unfathomable part. In other words, we all wear masks.

Chapter 21

▲ ■ ●

OCTOBER 5, 1982:
THE REVEREND AND MARTINIS

"Martinis or Manhattans?" Clark asked.

"We had Manhattans last time," I reminded him. "Let's do martinis." There was no need to tell Clark to put both olives and onions in mine. He knew how I liked them. In fact, I sometimes suspected that he knew more about me than I did.

The Very Reverend Clark A. Thompson had known me since I was a college freshman at Duke, more than twelve years now. We had met at a New Year's Eve party and took an instant liking to one another. He was bright, could talk in depth about nearly everything, and loved a good party—hardly what I expected from a man of the cloth. Over the years, Clark followed the development of my analytical mind, a pedagogical inclination, I suppose, given that he was the Starbuck Professor of Religion and Philosophy at Salem College, and the chaplain as well.[27] This made him a professional nurturer of mind and spirit. For me, his role was friend and confidant. When I moved to Winston-Salem to study at the medical center, Clark and I became nearby neighbors. Sunday afternoon visits, with cocktails, deep conversation, and walks around the neighborhood became our pleasant and inviolate ritual.

Now I was back in North Carolina as Clark's house guest. We took up our ritual of cocktails and conversation as if we had never paused during my absence. I had continuously briefed Clark on the status of synesthesia practically since the beginning, and I had just told him about my contract to write a textbook about it.

"An orgy of analysis," he said, pouring our drinks. "I can see your brain quivering now."

"Oh, stop it," I protested. "Believe me, it won't be a head trip. A book treatment gives me the chance to open synesthesia up to a whole range of topics."

"Such as?"

"Such as philosophy, and history, the art movement, creativity, things like that. It's an opportunity to explore the whole issue of subjective experience, maybe even say something about why science has been so hostile to it."

Clark raised his glass and grinned. "You've come full circle."

▲ ■ ●

What he meant was that he knew me back when. The teacher in him had quickly observed that I was a complete "head case." Worse, my innate analytic personality had been reinforced by twenty years of training in science and medicine. I reflexively analyzed whatever passed my way and firmly believed that the intellect could conquer everything through reason.

"You need an antidote to your incessant intellectualizing," Clark had once suggested, "something to put you in touch with the irrational side of your mind."

Such a suggestion was alien to my habitual way of thinking.

Clark suggested that I try Zen meditation, also known as "serene reflection" since its goal is to bring temporary quiet by stopping the internal dialogue of one's own mind. The party-loving preacher had actually journeyed to the mountains to study with Rōshi Jiyu-Kennett, the spiritual head of the Sōtō Zen Church in America.[28] Clark was most interested in religions that put their accent on the individual and meant something personally, rather than those that required rigid adherence to doctrine and rules. I had listened with interest when he recounted his Zen adventure, but when he suggested that meditation might be good for me, I ran the other way.

Being young and naïve, I had never considered that there might be more to the human mind than that rational part I was familiar with. It never occurred to me that a force to balance rationality existed, let alone that it might be a normal part of the human psyche. I was intrigued by Clark's calling Zen meditation "an antidote" to analysis. Yet despite the attraction I felt for dipping into this unknown, forbidden, and irrational ocean, I was so accustomed to intricacy that its simple method baffled me. I was supposed to sit in front of a blank wall, keep my eyes open, and do nothing! "Neither try to think nor try not to think," Clark instructed. "When the opposites arise the Buddha mind is lost. Just sitting with no deliberate thought is the important aspect of Zazen.[29] The internal dialogue will stop."

Always one to be in control (medical training strengthened that trait too), I protested that it was physiologically impossible for "nothing" to be going on in one's mind. Clark ignored my excuses and appeals to logical reasons why it was impossible to still the intellect.

"This isn't something that you question," he finally said, "because it doesn't have a rational answer. It's just something that you do. So do it if you want to. Or don't do it if you choose."

My own religious experience with Carmelite nuns in elementary school left me with a bitter taste for and distrust of anything religious. I could talk comfortably about such things and other inner experiences with Clark, however, because he did not fit my stereotypes of religious persons. If he had, I could never have accepted his challenge to my faith in reason.

He was a man of the world who accepted things as they were rather than subscribing to some ideal, but false, view of how things were supposed to be. Many of the Baptists and other thou-shalt-not fundamentalists around me insulated and distanced themselves from whatever opposed their world view. But Clark recognized the presence of evil, and immersed himself in the world while acknowledging its imperfections. When talking with him as an intellect, it was impossible not to simultaneously feel the great depth of his spiritual resonance. In fact, this side of him was a living embodiment of the old dilemma of how a rational person could believe in irrational things, such as faith. Clark's ability to balance his own rational and irrational sides made him the perfect foil to overcome my own resistance against letting go of my anchor in intellection.

Some part of me must have intuitively sensed the richness waiting in store, because after much persistence in front of that blank wall, I at last touched the still point. My cognitive mind was astonished that the internal dialogue really could be stopped, while the rest of me relished the sense of tranquility that accompanied this feat. It is a feeling that must be experienced to be understood, because it cannot be explained.

Clark's original point had really been very simple: too much analysis leaves you unbalanced. Fortunately, his friendship and our habit of probing conversation pushed me through a minor ordeal by which my own direct experience proved the existence of another part of the human mind. It also taught me its value. Years later, I was talking like a guru about inner knowledge and altered states of consciousness. I had come full circle indeed.

▲ ■ ●

"Why do you call synesthetes cognitive fossils?" Clark asked. "What does that mean exactly?"

"I know it seems a very romantic way of putting it," I said, "but frankly I want to emphasize how fundamental and absolutely basic an ability synesthesia represents."

"So, synesthetes epitomize not only what it means to be human, but, in a larger sense, mammalian as well?"

"Exactly. I don't mean that synesthesia is primitive, or that early humans possibly perceived the world synesthetically," I said. "People always seem to misunderstand that. I mean that it is closer to our biologic roots than ordinary experience is.

"Let me give you my television analogy," I offered. "A picture is what all of us see on the TV screen. Suppose someone were able to perceive the signals and make sense of them at a stage before they became a finished picture on the screen. That person would be just like a synesthete. Synesthetes are fundamentally closer to the ground of being alive and sentient."

"What a grand analogy!"

"I know it's hyperbolic," I said with a dismissive wave, "but I want to stress the larger view of who and what we are. When you can do that, so many of our assumptions fall away. We are more than what we appear to be and we know more than we think we do."

Clark laughed. "I know I encouraged you to consider non-rational modes of thinking, Rick, but you almost sound like a mystic now."

"Oh, I don't think so," I protested. "My point is simple. We need to be in touch with our irrational and emotional sides as well as our rational one. I don't mean like Rousseau, but like D. H. Lawrence or E. M. Forster, writers who suggested how modern life unbalances individuals in such wonderful scenes as those in which people walk naked at night through the forest, brushing the pine trees up against their skin, that sort of thing. The unclothing is metaphoric, isn't it, urging us to cast off the confines of reason so we can reclaim more direct experiences. You find the same message in American Indians and in primitive cultures, and there it's even more blunt. Surely I'm not really saying anything new, am I?"

"No," agreed Clark. "Religion is full of symbols of casting off external things to find internal truth. Unfortunately, it is the point of view so often shot down by proponents of objectivity, technology, and the scientific method, as if nothing can be known or accepted as real unless it can be measured with an instrument that cannot 'lie' as our senses can. It always insists that experience needs to be verified by the instruments of science because the mind is too unreliable to arrive at truth."

"Technology refutes direct experience in favor of abstract ideas," I agreed. "Take time, for example. Not too long ago, people acted according to biological and natural time scales. They ate when they were hungry and slept when it got dark, not when an abstract model reached a given state, as we do today when the hands of a clock reach a certain position. They took their cues from the sun, from the conditions of crops, from the behavior of animals, and from their own bodies. The appearance of the clock replaced direct experiences with a disembodied machine. Earlier people experienced time as something personal and direct, as a cycle of recurring events, not as a succession of abstract moments that we now 'know it to be' thanks to science."[30]

"And what does this have to do with synesthetes as cognitive fossils?" Clark asked.

"Well, synesthesia is the most immediate and direct kind of experience I've ever encountered. It is sensual and concrete, not some intellectualized concept pregnant with meaning. It emphasizes limbic processes which break through to consciousness. It's about feeling and being, something more immediate than analyzing what is happening and talking about it. The immediacy and simplicity of it go directly to the heart of the matter."

Clark mulled over what I had said. "Synesthesia does seem to have a lot in common with the eureka moment of insight, mystic experiences and religious rapture, what we call noëtic experiences. You may have latched onto something big, Rick."

"How so?" I asked.

"It's an old argument in philosophy and religion that reason is not the only way to truth. Reality is not restricted solely to what is given by sense experience. It would be refreshing to hear a scientist articulate such things," he smiled. "Basically, you're confirming old notions that the conscious mind does not direct everything."

I waved my empty martini glass at him. "I think my unconscious mind wants a refill," I said. "And some more olives."

"Marvelous idea," smiled Clark. He stood up and collected our glasses. The room had gone gray as dusk approached. "Why don't we move out to the porch," he suggested. "It's going to be a mild night."

Clark banged around in the kitchen, fixing our refills and some hors d'œuvres. I went outside and settled into the porch swing in time to savor the horizon, which the last rays had colored a surrealistic orange. Two blonde girls approached on their bicycles, one shouting at the other.

"Wait till I get you, Sally Anne!" shouted the smaller girl, pedaling furiously behind her provocateur. "Hold up, here! You hold up!" she hollered as they rode by. Nobody can throw a fit like a Southern girl-child can, I thought. I wished I could be at the finish line to see how Sally Anne acquitted herself.

The indignant shrieks of the little miss died down, blending into the evening soundscape as she and her sister receded down the road. What an unspoken skill riding a bike is. I thought back to my own scrapes, falls, and frustrations and how I now took riding a bike for granted, just like I did so many other skills I had once struggled to master. We never have to consciously think about a skill once we learn it. This is a maxim of human experience. "Motor memory" is the neurological term for behavior that becomes automatic after learning. It is a commonplace observation that any performance, like playing tennis, or the piano, even giving a speech, is impeded if you try to think about it. Thinking about what you are doing degrades your ability to do it.

Psychologist Karl Lashley said as much during the famous Hixon Symposium at the California Institute of Technology in 1948.[31] At that time, the method favored by scientists interested in the brain was introspection, which is conscious self-reflection. The dominant theoretical framework was a linear chain between a stimulus and one's response to it known as behaviorism. Lashley challenged both ideas by pointing out that the sequence of actions in complex skills such as speaking, playing tennis, or improvising jazz unfold so rapidly that there is no time for feedback, no way in which the next step in the supposed "chain" can be based on the previous one. And yet there are millions of people who can successfully improvise jazz at the piano.

Lashley suggested that the overall form of behavior was not imposed from without, by stimuli, but that its organization emanated from within. The idea that sequences of behavior, which is what any skill is, are planned and organized in advance sounds very much like the later work of Kornhüber and his followers. What we know of as our conscious, rational self is not in control; some other part of us is. Moreover, this unfathomable part is capable of producing some great behavior, which is all the wonderful, irrational, and interesting stuff that humans do. Disciples of AI keep trying to reduce the human mind to mathematical statements of formal logic, apparently oblivious to the fact that formal problems are a tiny subset of those that face human beings. Even then, formal problems are of interest mostly to mathematicians, physicists, and engineers—precisely those people who are trying to build AI—not your average person, such as livid little girls riding bicycles.

Ordinary people are faced with all sorts of novel problems every day, situations that produce disorder in their lives. Without knowing how, they manage to solve problems they have not faced before and somehow bring a creative solution to bear on them in a way that brings order to a previously chaotic situation. People do not resort to rational rules at such times. They just act. This shows that our *ideas* about everyday experience are flotsam on top of the *actual experience*.

If, for example, we try to explain the grammar of English, we would soon see that a six-year-old has more grammar in her head than is contained in all the rules we can possibly write down. The linguist Noam Chomsky demonstrated that natural languages have innate, complex, subterranean structures that become disrupted the moment you start to bring consciousness to bear on them.

Elementary school children used to be taught how to parse sentences in the course of learning grammar. Because of this, experts in the 1960s thought that they could build a computer, give it a dictionary and a parser, and let it snappily translate sentences from one language to another. Were they in for a surprise! Language turned out to be far more complex than the grammar found in textbooks. And yet it is routinely surpassed by what a six-year-old has in her head.

The objective world view refutes direct experience in favor of abstractions. Yet ordinary experience shows us over and over again what a conceit the objective view is. Bringing reason to bear on what we are doing often interferes with it. Rational logic does not change the baby's diaper, find the file you are looking for, or drive you to work.

▲■●

Clark maneuvered through the screen door carrying a tray. He set the plates down on the wicker table. "You look deep in thought," he said, handing me my glass.

"This may sound strange since I've spent my career steeped in analysis," I admitted, "but I've come to the ironic conclusion that rationality is overrated. It would be hard convincing others, particularly my colleagues, but it seems clear to me now that humans have always been, foremost, emotional creatures."

"That's a sweeping statement," Clark said, drawing up a wicker chair. "Do you really want to argue it?"

"The follies of humankind argue it for me," I said. "Persons who believe they act rationally are experts at deluding themselves. What they are really doing is rationalizing their emotions. Our psychic closeness to contemporary events blinds us to this reality, so the fact that human motivations are irrational is easier to see retrospectively, in the lessons of history. Didn't the 'noble savage' from the Age of Reason turn out to be more savage than noble?" I asked. "Can the savagery of religious and ethnic wars support the empty contention that such disputes are based on rational facts?"

"The situation in the Middle East and across Eastern Europe is complex," Clark interjected.

"I agree. But that complexity leads back to all the intellectual somersaults and political posturing that are necessary to rationalize intense hatred. Can you possibly imagine a politician saying 'We hate you and we will kill you'? What they spout instead are abstractions about diplomacy, ethnic majority, and negotiations over material things, all of which are displacements of the underlying hatred."

"Politicians have sold their potential for values that are inscrutable to me," Clark said, "but I cannot believe that all political behavior is best explained by self-interest, greed, and lust for power."

"Duplicity is the government's trademark," I countered. "The thought of doing the right thing instead of what was expedient actually crossed a congressman's mind once, but he mistook it for a nightmare."

"Did you want me to listen to you seriously?"

"All right, if politicians and governments are not rational, then how about corporations, lawyers, investment bankers, or even the clergy?" I continued. "What motivates journalists and television programmers? Tell me, if you can, who qualifies as a rational being."

"I get your point," Clark said.

I knew that Clark had taught a course about altered states of consciousness, from both religious and psychologic points of view. He was acquainted with split-brain research. "Do you remember the hullabaloo about how schools had neglected educating the right side of students' brains?" I asked.

"Oh, the battle for the touchy-feely curriculum," Clark exclaimed. "Indeed I remember."

Shortly after knowledge of split-brain research entered popular culture, a fad for right-brain education swept the nation. Parents, educators, and institutions sought to turn schools upside down, accusing them of systematically neglecting the holistic, intuitive, and artistic aspects of the right brain through generations of narrow-minded focus on the verbal, logical, and analytical skills of the left hemisphere. People seriously believed that the three R's of reading, 'riting and 'rithmetic had been a tragic mistake.

"The critics claimed that traditional education was turning children into unfeeling, analytic automatons," I pointed out. "Some colleges actually planned sweeping changes despite the argument being patently absurd."[32]

"The whole thing was a fiasco" Clark said, "but I'm not so sure the premise was absurd."

"I maintain it was," I insisted. "It oversimplified the research and based grand schemes on its misappropriation. Even assuming its truth at face value, we would be a country of cold Mr. Spocks, exclusively verbal, logical, and analytical. Yet wherever you look for evidence, you find a nation populated with everything *but* verbal, logical, and analytical people."

"You and I being exceptions, of course," Clark chuckled.

"I have a point to press," I answered. "For more than a decade, American students have scored at the bottom when tested against those of other nations in math, geography, science, world affairs, and the ability to read and reason. One would hardly expect such dismal performance if we had been grinding out generations of lopsidedly analytical students. The premise of the complaint was itself an irrational, emotional response to new information," I argued. "Saying that humans are rational is just as preposterous as saying that we've neglected to educate the right hemisphere. People don't think about what they're saying."

Clark drained his glass. "Then why is it a commonplace assumption that reason is the hallmark of humanity?"

"Because the contemporary view of how the brain is organized has not yet disseminated into popular culture. Nobody's written books about it like they have for quantum physics or space travel. Current biology and anatomy show that the relationship of the cortex to the limbic brain is the reverse of what we had always assumed it to be. Sometimes, Clark, you just have to stand on your head to see things clearly."

"And you need to do so regarding rationality?"

"Yes. I'd like people to recognize how serious a role non-rational processes play in their lives." I sighed, stopping to scoop up an hors d'œuvre. "Of course my colleagues didn't even want to hear about such things, so I have no illusions that it will be difficult convincing others."

"It doesn't matter what field you work in, Rick. You'll always find closed minds, and any time you go against the party line you'll feel resistance. There wasn't anything special about their reaction," Clark assured me.

"Why did they feel so threatened?" I asked.

"All people with narrow world views are threatened by change. You're questioning why we routinely dismiss direct experience and refuse to let things present themselves as they are. You're suggesting that we are more than individuals with many facets. Perhaps we possess multiple minds. How do you expect people with no vision to react?"

"I feel it's important to advance the issue," I answered. "I've been stunned at how often everyday assumptions about the way things are turn out to be wrong, or even backwards. You've got to invert preconceived notions to make sense of them again. That's what I mean about standing on your head."

There was a comfortable silence between us. I rocked slightly in the porch swing, just enough to make the chains creak. "I've lived most of my life insisting that everything that exists can be objectively described, explained, and controlled by the scientific method," I said after a while. "Perhaps I'm disillusioned at seeing the limitations of analysis so clearly now."

"I doubt that," Clark replied with pointed emphasis.

I stopped rocking. "What do you mean?"

"I mean that you will do what you always do," Clark replied calmly. "You'll find a way to integrate what you have learned."

Clark's talent for hitting the mark was as illuminating as it was uncanny. "The fact that synesthesia is known to the smallest fraction of humanity does not mean that it is insignificant," I said. "I've never had such an experience, but I can still marvel at what my brain is capable of knowing. And I have to assume that it is capable of much more than what I'm familiar with. I feel drawn to explore the possibility that all people may be capable of creating their own spectacular universes."

Night had come. We listened to the crickets as we sat quietly on the porch. At last Clark spoke. "That's quite a vision, Rick. It has broad implications. Most altered states of consciousness do oblige reality to take on a new perspective for the rest of us. I think anything that challenges people to explore their inner life, and hopefully expand it, is valid."

"You think so?"

"Teaching religion, as well as ministering it, tells me that people are always intrigued by the possibility of stepping into a larger world that has been waiting for them all along. Usually they hesitate to talk about it. The taboo against discussing inner matters is especially acute for people who either had unpleasant religious experiences early in life or who think of themselves as analytic and objective. People often need permission to explore different aspects of their lives. Maybe an objective doctor explaining synesthesia will give them the encouragement they need to explore their own inner life."

Part Two

▲ ■ ●

ESSAYS ON THE
PRIMACY OF EMOTION

▲ ■ ●

A Note on the Essays to Follow

The following essays develop my theme that what you know as your conscious mind is not the arbiter of what is real and true. It is not even the agent in the driver's seat that we unthinkingly call the self.

This idea is not original. For example, in a book called *Minimal Rationality*, philosopher Christopher Cherniak concludes that humans are a little bit rational, but not very much so. "The pervasively and tacitly assumed conception of rationality in philosophy is so idealized that it cannot apply in an interesting way to actual human beings."[1]

As Clark would have said more simply, we need an antidote.

The carnival magician who pretended to be the Wizard of Oz shouted, "Pay no attention to that little man behind the curtain!" He wanted all attention focused on the powerful apparition he had created. Similarly, Western cultures identify self with the seemingly powerful, outwardly projected rational mind. However, the truth is that the irrational part of ourselves, the emotional, heuristic mind, is the man behind the curtain and the one really in charge.

The essays that follow spin out a few ramifications of this idea. They are essays in the original sense of that word—trials or attempts. They are meant to be suggestive to the reader, to provoke both thought and feeling, and not to be the final word on the subjects that they treat.

▲ ■ ●

Chapter 1

▲ ■ ●

THE ANTHROPIC PRINCIPLE

The topic of consciousness was once the exclusive purview of philosophers. While efforts to relate the ephemeral mind to the solid brain date back to antiquity, it was not until the nineteenth century that biologists got deeply hooked by the problem. Taking coma as an example of un-consciousness, they simplified the philosophical muddle by equating consciousness with wakefulness and awareness. Instead of asking "What is consciousness?" they were more interested in its anatomy, or "Where is consciousness?" In the twentieth century, a small group of abstract thinkers became preoccupied by the concept of a conscious machine, the idea of artificial intelligence.

Now everybody wants to get into the consciousness act. I am referring to scientists rather than Californians or New Age devotees. Even scientists from non-biologic fields that never considered it before are passionately interested in consciousness nowadays, believing it to be widely important. Nobel Prize winning Immunologist Gerald Edelman has written three books about consciousness, for example, and mathematician Roger Penrose made a big splash with *The Emperor's New Mind*.[2]

A recent proposition called the "anthropic cosmological principle" suggests that mind, like matter, is a fundamental property of the universe. Such a premise raises cardinal questions about the nature of reality and brings us face to face with an ancient and essentially spiritual issue. If you believe in what is called the *weak anthropic princi-*

ple, then you hold a view of human supremacy which says that the universe is the way it is because we are here. Since we exist as thinking beings, then the physical properties of the cosmos must be such as to allow our kind of being to have evolved. The evolution of life to include brains and self-aware minds is part of the weak anthropic principle. Therefore, it assigns us an essential role in the very creation and unfolding of the universe. Perhaps our will is even its central force. The *strong anthropic principle,* on the other hand, says that consciousness is inherent in the cosmos and that human minds are only a single instance of its many possible manifestations.

At the moment we can not prove either position, so each remains a matter of faith. However, each has different consequences depending on which one you believe. Pushed to its extreme, the weak anthropic principle concludes that humans are meant not to live in accord with nature, but to conquer it. A weak anthropic believer assumes humans to be at the summit of evolution and therefore holds, as Martin Buber would say, an I-It relationship between self and everything else. Tools are one objectification of this I-It relationship. Both a Neanderthal knife fashioned by flaking a stone and today's space shuttle are tools, instruments that by definition conquer nature by turning it into human service. The depth of the I-It attitude of course varies from person to person, with extremists feeling entitled to cut down forests, drive species to extinction, and squander natural resources because, they believe, the universe exists just for them.

Believers in the strong anthropic principle, on the other hand, feel that consciousness is not solely ours. Other planetary inhabitants and the universe itself may have a soul, they say. Jesuit theologian and paleontologist Pierre Teilhard de Chardin suggests a mind principle, a spiritual inevitability hovering over the process of evolution: he proposed that we would not have brains capable of spiritual experiences if there were no sprituality in the universe. Strong anthropic believers have an I-Thou relationship between self and non-self, sensitive to the implications of their interactions with other species and the environment as a whole.[3]

If the cosmos *is* conscious, then our scientific notions of objective reality come directly in contact with dimensions reaching beyond what we understand as factual. This is the realm of religion. On the face of it, the collision tends to make people uncomfortable, although what is probably common to all religions is nothing more than the claim that something more exists than what we factually receive through our senses. Because religion concerns that which we cannot verify directly or with technology, we believed for a long time that science and religion were incommensurable. Now we understand better that some questions are simply not answerable by the scientific method.

We are reaching a point in human history where people are increasingly replacing self-centeredness with sensitivity to the Other. We have rediscovered a new respect for the earth in a way reminiscent of indigenous cultures, who were not so removed from direct experience as we are today. There is growing acknowledgment of the intelligence of other species, from ants to whales, even though theirs are qualitatively different intelligences than ours. Technology permits us to explore undiscovered country in the comfort of our living rooms. Images from the ocean floor to the edge of our solar system show the cosmos to be more wonderful than anyone could have imagined. Ironically, it may be technology, that relentless conqueror of nature, which has sensitized viewers to the possibility that we are a part of nature rather than apart from it. What we do to the web of life we do to ourselves.

One can perceive this concern in the Gaia concept, for example, which sees the planet itself as a conscious, self-regulating macro-organism. I will set aside criticisms of whether the Gaia concept is New Age nonsense because what I find noteworthy is not so much the question of its plausibility as the emerging human preoccupation with realms larger than the self. Such global attitudes indicate that people are increasingly aligning themselves with the strong anthropic principle. But why?

I proposed that subjective experience and emotion form the quintessence[4] that is uniquely human. Perhaps individuals just intuitively feel it is the right thing to do. An increased sensitivity to the Other and a belief in realms larger than oneself is not a rational attitude that people vote on or deliberately decide to adopt. It is an emotional one.

Chapter 2

▲ ■ ●

FREE LUNCH AND IMAGINATION

The human limbic system enhances cortical processes, and by extension reasoning, because it is surprisingly energy efficient. Its ability to reduce entropy,* act on incomplete information, and create order from a continuous and incoherent sensory stream is what gives us an æsthetic capacity. Without emotion, our behavior would be altogether predictable and unimaginative. One can, for instance, demonstrate that human limbic activity drops out when newly learned actions become mere habits.

Up to the level of reptiles, brains evolved into increasingly complex neural systems that were still hardwired, thus yielding totally predictable behavior despite their increased complexity. The limbic system, which first appeared in early mammals, is most richly redundant in humans. By sharing structures and pathways for such different functions as attention, memory, emotion, and consciousness it is able to act on incomplete information. Its ability to determine valence and salience (value and relevance) yields a more flexible and intelligent creature, one whose behavior is unpredictable and creative.

---

*Entropy is the measure of that part of a system's energy which is not available to do work. More commonly, we use it as a measure of disorder. A broken glass and a scrambled egg have more entropy, for example, than their intact counterparts do.

Disorder tends to increase if things are left to themselves. (You have only to not clean house for a week to prove this.) But you can create order from disorder (e.g., you can clean the house) by expending effort and energy. In doing so, however, you decrease the overall amount of ordered energy available. This idea is the second law of thermodynamics: the entropy of irreversible processes (breaking glasses or scrambling eggs) of enclosed systems always increases.

The limbic brain's use of common structures for different functions *in the same system* is not an accident, but the optimal development of a means to evaluate incomplete information and initiate actions at the least energy cost. I described how assumptions of "bigger is better" failed to distinguish human brains from those of other species. What appears unique about the human brain is its high energy consumption coupled with fantastic efficiency in producing mental work. The brains of rats and dogs, for example, consume five percent of total body energy, monkey brains use ten percent, and human brains expend a whopping twenty-five percent, far greater than expected for its relative size.

Authorities on brain metabolism tell us that, remarkably, hardly any energy is used for mental work. Instead, nearly all the energy is consumed for housekeeping, mainly the pumping of sodium whereby electrical charges are maintained within nerve cells. The energy cost of mental work itself is minuscule compared to that consumed in just keeping up the physical structure. Getting something for practically nothing makes mental work almost the only free lunch in the universe.

Mentation that needed much more additional energy would produce a brain that consumed so great a proportion of bodily fuel that life would probably not be possible. It is the organization of emotion that permits human cognition to occur with only slightly more energy use than is required to sustain the brain's physical structure. As incredible as this free lunch sounds, it does comply with Planck's principle of least action. In 1922, Max Planck, the Nobel laureate physicist, formulated a rule to predict which of several alternate paths a given event would take. Planck's principle says that among all possible paths, the one actually taken is always the one that uses the least energy. Such a general principle finds itself exquisitely expressed in the efficiency of the human brain.

All creatures maintain their lives by consuming organic matter in the food chain, a process that ultimately goes all the way back to the conversion of solar energy by photosynthesis in plants. The energy cost of each conversion in the food chain is huge, and the exceptional efficiency of the human brain is again noteworthy in this context. With injuries such as trauma, burns, surgery, or infection, the body becomes hypermetabolic, consuming more nitrogen and other essential nutrients. Strikingly, when the brain alone is injured, the body's hypermetabolism increases proportionately to a greater extent and the rest of the body will be starved if necessary to protect the brain.

However, there is a mental cost for this inefficient hypermetabolism. The greater the body's hypermetabolism, the less efficient cerebral energy consumption becomes, resulting in greater cognitive impairment. As cognition returns to normal, so does the efficiency of cerebral energy utilization. The point is that energy efficiency is a product of how the limbic brain is organized, and this in turn influences the analytic processes of the cortex. In determining salience and valence, what we might call *qualitatively significant information*, the emotional brain acts like a valve, regulating the flow of nervous information throughout the body, integrating both the direct wiring and volume transmission systems.

As if these facts were not impressive enough, there are also rather advanced proposals, along with the calculations of irreversible thermodynamics to back them up, suggesting that all life forms, but particularly brains, play a large role in slowing down the rate of entropy increase and the degradation of energy in the universe.[5] Such profound possibilities suggest that we should direct our efforts not toward controlling our emotions but toward gaining better insight into them and into the fundamental role they play in our lives. For such reasons as these, I have been more and more drawn to the assertion that emotion frees us from the tyranny of predictable, reptilian-like thought and behavior.

I indicated that were it not for emotion, our mentation would be predictable and unimaginative. The ability to pluck qualitatively salient information from the passing stream and to act efficiently on fragmentary information is what leads to imagination and an æsthetic capacity. Intuition, for example, is the expression of a decision based on the efficient use of partial information. Humans excel at this, which is a blessing given that human thinking is neither inherently logical nor clear.

Because the anatomy of emotion is also partially the anatomy of memory, increased clarity comes from the capacity to look at and to remember previous actions. The reasons we develop when talking to ourselves become clearer as we gather and retain more knowledge about our motivations, the way we make decisions, and how we rationalize our actions. Seemingly irrational things like contradictions are a natural part of this process. We develop dichotomies, such as good versus evil, to clarify our thoughts. It is probably impossible to understand anything without making such polarities. Some polarities are true and have a physical basis, like positive and negative.

Others are elusive and resolved only by perseverance. When we eventually fathom a linkage we call it *insight*.

Some people are blessed with an intuitive grasp of relationships between huge numbers of variables without having to "reason" their way through them. Such a person was Ramanujan, the Indian boy who developed mathematical theorems he had invented himself. Because the proofs were so obvious to him, he only wrote down the theorems. The British mathematician Hardy declared that Ramanujan's formulas had to be true because no cheat could have fabricated work so sophisticated. While few of us possess intuition this grand, we all possess this uniquely human type of creative thinking that brings order to a jumble of variables and somehow gives meaningful pleasure.

When faced with a totally new problem in a new context, we somehow come up with a creative solution. This happens all the time in ordinary human situations like raising children or trying to get someone to notice us. We solve myriad daily problems that involve our relationships with the world and with other people. It would be difficult for a logical machine following rules to deal with such problems, no matter how detailed the rules were. If we could define emotions in terms of rules, then perhaps we could stick them into such a machine and make it intelligent in the sense hoped for by AI enthusiasts. But emotions cannot be set down in formal terms; they can only be understood by living life and feeling our way through it.

Creative people who do original work do it with an emotional charge, and the greater the charge, the better they seem to work. This kind of emotional tone seems to be the guiding force in any new creation. Emotion not only makes our brain so efficient but also gives us our intuitive sense of what is correct and what goes together. This is of course the capacity for an æsthetic, a sense of what is beautiful and not beautiful. Without such a capacity we probably would not have higher realms of creative thought, such as literature, architecture, or mathematics.

While pointing out the overlap between emotion and memory, I want to emphasize that memory is not simply a fixed look-up table. It too is a *creative process* during which the state of the brain's electrical fields change. The sensory cortices generate a distinct pattern for each act of recognition and recall, with no two ever exactly the same. They are close enough to cause the illusion that we understand

and have seen the event before, although this is never quite true. Each time we recall something it comes tainted with the circumstances of the recall. When it is recalled again, it carries with it a new kind of baggage, and so on. So each act of recognition and recall is a fresh, creative process and not merely a retrieval of some fixed item from storage.

Furthermore, persons, objects, and events are not perceived in their entirety but only by those aspects which are, have been, or can be experienced and acted upon by an observer. An example of this fragmentary nature of perception is found in a mundane object such as a disposable plastic cup. Everyone knows that you drink from it. But we can comment on little else—for example its tensile strength, translucency, thermal coefficient, chemical composition, or what is stamped on its bottom. A physical universe is contained in that cup by such an analysis. Yet all we really know about it is *what you do with it*. This limited aspect of knowing is peculiar to humans, the observers and manipulators. All that we can know about anything outside ourselves is what the brain creates from raw sensory fragments, which were actively sought by the limbic brain in the first place as salient chunks of information.

This view of perception and memory does not appeal to the idealism of philosophers because it speaks to the *limits of our knowledge and what we really know*: namely, conscious knowing is restricted by the possible interactions we have with events and things. Conscious knowing is based in direct, hands-on experience.

Put in a more familiar context, artists and creative writers look at the world in a certain way. It is the same world that everyone else sees, but seen differently. Contemporary people often call artists weird because they do not seem to be seeing the same things that the majority sees. It is critical to realize that the sensory gateways that feed into the brain establish their own conditions for the creation of images and knowledge. Artistic giants knew full well that their visions were not shared by most people. Even when persecuted or abandoned because of their vision, artists persist. That is all they can do because their visions are their reality, and for many of us they subsequently become our reality when we experience their art.

Chapter 3

▲ ■ ●

CONSCIOUSNESS IS A TYPE
OF EMOTION

What is consciousness? Anyone who has surveyed the philosophy of mind or biological explorations of consciousness is aware that consciousness is most often identified with reason. I would like to propose a radical alternative: that consciousness is a type of emotion.

Consciousness is firmly tied to emotional drive and goal-directed behavior. We are interested not just in whether a wakeful state or self-awareness is present (typical definitions of consciousness), but in whether a creature is capable of purposeful action. Studies of natural and deliberate brain lesions clearly indicate that the cortex is not necessary for propulsive, teleological behavior. In fact, monkeys who have had their cortex surgically removed can barely be distinguished from their normal cage mates, a profoundly counter-intuitive observation, surely. Animals who have had their cortex and even important motor structures below the cortex removed still show purposeful behavior as long as their limbic system remains intact.

This may be counter-intuitive but is not really surprising, because the limbic system contains tissue central to the formation of memory, and goal-directed behavior requires continuity over time. Something that reacts instantaneously to a given stimulus will not do. You need some sort of register, or memory, to direct behavior from within. The hippocampus of the limbic system is ideal to couple memory with motor readouts from various parts of the brain, thus enabling purposeful, goal-directed behavior.

I am not saying that the hippocampus is the seat of consciousness.

It is no more so than the cortex is. Maybe we should stop trying to put a finger on consciousness. *Maybe it is not a thing, but a relationship between oneself and the external world.* Just as gravity describes a relationship between masses, perhaps "mind," "consciousness," and similar terms refer to relationships between an organism and its environment. There is good reason to conceive of this relation as an emotional state: the calm but indescribable one that we are in most of the time. In other words, consciousness is a type of emotion. Here is the reasoning.

Anybody familiar with engineering knows that you cannot have an "up" state and a "down" state with nothing in between. And yet we typically describe emotions in terms of high and low, never neutral. When it comes to emotion, this in-between state always seems to be overlooked. We are *constantly* in an emotional state that can go up or down, depending on circumstances. It can also remain chronically down, so that we experience clinical depression, or chronically up, as in mania and delirium. It is impossible not to be in some emotional state at every moment. Most often we are not in one of the up or down states considered typical of emotion but are in this in-between state that happens to be fairly calm.

Its exertion of a constant force signifies that emotion influences everything. An impressive example of its influence is seen in epilepsy. Persons with epilepsy are always afraid of seizures and attempt to control all excitement because they know that emotional stress can trigger seizures. Nonetheless, patients have seizures despite maximum exertion of will because they do have an underlying medical problem and there is only so much that self-will can achieve. However, their failure to prevent all seizures leads to self-recrimination, frustration, depression, withdrawal, more frustration, and soon more seizures. They are in a no-win situation because the underlying emotional state constantly acts on their entire nervous system.

Specially dedicated seizure units at research institutions vividly highlight this relationship. Persons who have thirty to forty seizures a day come into a ward that is devoted entirely to understanding epilepsy. They are taken off all medication and kept in a room where injury is unlikely. Constant video monitoring and EEG recording collects data on exactly what kind of seizures they have. Remarkably, their seizures often stop for weeks because they have come to a place where they believe they will be better. There is such a tremendous relief of stress from being in the hands of perceived healers that with simple faith their condition is better for a while. Of course this temporary change in environment cannot cure the underlying problem, but it does demonstrate the tremendous impact that emotion can have.

The idea that consciousness is a type of emotion was recently proposed by Ayub Ommaya, an engineer, philosopher, and former head of neurosurgery at NIH. His proposal that the brain's emotional core controls the rest of the nervous system arose from powerful contradictions in existing theories that equated consciousness with reason. I discussed earlier the explicit anatomic and physiologic reasons why it has taken until now to understand that the limbic system, which is intermediate in the brain's evolution, regulates the cortex more strongly than the cortex regulates the limbic system.

Dr. Ommaya suggests that emotion monitors the state of conditions at any given moment, providing a continuous index of what is going on. He calls the brain's emotional organization "a fundamental strategy in evolution which enables increased efficiency in energy utilization for success in any ecological niche. This evolutionary strategy is first noted in the late reptiles, and developed dramatically in birds, mammals, and, most of all, humans. The mechanism for this strategy is found in the anatomy and physiology of the limbic system with its high degree of reciprocal connections within itself and with all other levels of the brain."[6]

This formulation restates many of the ideas we clarified in Part I and also draws heavily on the brain's ability to produce mental work at the least energy cost. In evolving into the structure that it is today, the brain still had to conform to Planck's least energy principle, and many of our higher mental functions are a consequence of that conformity. We explored in Part I how the expression of synesthesia is dependent on the limbic system and is itself the basis for more abstract cross-modal associations such as language. Said differently, we concluded that language is the consequence of more fundamental cross-modal associations such as synesthesia. The more universal articulation of this arrangement sees consciousness, language, and higher mental functions as the *consequences of our ability to express emotion*. Emotion is fundamental to mind and what we call consciousness.

Chapter 4

THE LIMITS OF
ARTIFICIAL INTELLIGENCE

In the past, defenders of humanity pointed out that artificial intelligence (AI) machines were incapable of pleasure, desires, or hope—what philosophy calls "qualia," the feelings we refer to in describing humans as different from machines. AI proponents maintained that thinking was nothing more than a set of formal rules, and qualia were therefore nice but unnecessary. AI's standard rebuttal asked humanists to prove why emotions were essential for thinking. Hoist by their own petard, engineers who build neural networks have recently discovered that a machine's performance can be drastically improved by giving it something like emotion. Here is what led to this reversal.

Practitioners of AI have always taken the separation of mind from brain as a given, believing mind to be an abstract program that can be instantiated on any machine capable of running it. They believe that "understanding" the mind is really a *technical* problem of reducing it to a series of formal logic statements, and that the scope of human experience is in fact so reducible. The suggestions that human psychology might not be separable from its biology is an odious thought to the AI camp, an attitude reminiscent of early Vatican "scientists" who refused even to look through Galileo's telescope. AI proponents do not want the inconvenience of having to learn biology, nor do they want to hear why human logic might not be separable from emotion.

The approach that sees the mind as software—some abstract, disembodied knowledge—is extremely attractive to people who think theoretically precisely because it liberates them from having to learn the biological complexities of neural tissue. Yet there are two reasons, one mechanical the other moral, why we should be suspicious of claims that the brain can be modeled faithfully by this approach.

On the mechanical front, the distinction between software and hardware has become essentially meaningless. Engineers of neural networks invent mathematical designs that can either be run as programs or translated directly into computer chips if need arises. They extrapolate this practice to mind, saying "define the program, and running it on a machine will fall into place." This confidence seems overblown given our failure at replicating far simpler biological parts. In 1970, for example, engineers at Pennsylvania State University enthusiastically promised that an implantable, artificial heart pump could be rapidly developed and be ready for routine surgical use by 1975. After twenty years of effort and billions of dollars, this goal remains elusive. Other devices, such as artificial joints and kidneys, are likewise nowhere as sophisticated as the biologic counterparts they claim to replace. Yet listen to Herbert Simon, possibly the most influential elder of AI, assert the following in 1958:

> There are now in the world machines that think, that learn, and that create. Moreover, their ability to do these things is going to increase rapidly until—in the visible future—the range of problems they can handle will be coextensive with the range to which the human mind has been applied.[7]

Nearly forty years since this sweeping declaration, AI has not produced any general principles about thinking. What are called expert systems have indeed been impressively successful in *highly specific tasks* such as medical diagnosis or financial analysis. But a machine must succeed in *general* situations before we can call it intelligent. We still await the bare-bones model, the vanilla AI without qualia. We would hardly call a calculator that added some numbers correctly but not others a formalization of arithmetic. Likewise, we cannot call the successful replication of a few specific reasoning tasks a formalization of general thinking.

This has not dissuaded engineers who build neural networks, which are modeled on some of the hardwired circuits of the human brain (AI has yet to address volume transmission). They cite the brain as living proof that hardware consisting of analog-distributed circuits can act as a controller. In fact, they have successfully installed neural networks to run dangerous chemical plants, manage inventory, and oversee other industrial tasks. This success in highly limited settings has only made them more certain that replicating the spectrum of human thought is not far behind. Back in 1960, Herbert Simon boasted that "duplicating the problem-solving and information-handling capabilities of the brain is not far off; it would be surprising if it were not accomplished within the next decade."[8]

The real surprise turned out to be the discovery that while existing neural networks do work to a limited extent, giving them something like emotion makes them perform even better. This makes it worth explaining the three parts of a neural control network. They are (1) a model (of an assembly line, say, or whatever domain the network is to manage), (2) an action system (for welding, putting bottles in cartons, etc.), and (3) an adaptive critic.

The adaptive critic checks the result of every action taken against the model and feeds back to the action system telling it whether it did the right thing or not. A yea-or-nay critic based on a single outcome measure works fine for simple, monotonous tasks such as assembly-line work. But it fails at tasks with numerous variables, which is what humans routinely deal with. If you are twiddling dozens of dials and the critic can only say that you are doing well or lousy, how can you know which dial led to the action being criticized? The problem becomes exponentially worse as the number of variables increase.

There turns out to be only one fundamental principle on which to build a critic that addresses the problem of too many variables. Paul Werbos of the National Science Foundation credits the basic idea to Freud's model of how neurons interact biologically.[9] Freud envisioned an emotional charge (called cathexis, or psychic energy) that drives every action we take. The network critic assumes this role, generating an evaluation of the ongoing action. Freud said that every object carries an emotional charge, and if A causes B, then a corresponding and proportional flow of emotional charge must propagate backwards from B to A. This led to Werbos's idea of back propagation, which is an essential feature of modern networks.

Werbos says that each component of his network has a counterpart in the human brain. He sees the brainstem and cerebellum as the action component for motor output; our objective model of the world in the cerebral cortex; and the critic in the limbic system. The limbic brain and cortex perform different functions. The limbic system provides an *evaluation* of what one is doing, generating the emotional charge that Freud talked about. The cortex contains a representation of reality. Memory serves as a check on both the model and the critic.

Aside from the fact that incorporating a model of emotion makes networks work better, it is also surprising how closely the engineer's ideal design for a sophisticated controller matches that of human biology. Engineers say that an effective critic must have high-speed recurrence. The limbic system performs calculations at an internal cycles-per-second rate of 400 Hz but is governed by a much slower outer clock of 5 Hz, the rate of the theta rhythm. In other words, a high-speed calculator is embedded in a low-speed clock.

The cortex also performs high-speed modular calculations governed by a low-speed clock of 10 Hz, the frequency of the alpha rhythm. This 2:1 ratio is what engineers require to adapt a critic. You need to hold, store, and reëvaluate in a way that makes the cycle time of the critic twice as long as the model's cycle. This 2:1 ratio exists between the limbic brain's evaluation and the update of the model in the cortex in the following way: the state of the world is pumped into the cortex and an evaluation comes out one fifth of a second later, yet elements *inside* the limbic system are cycling furiously 400 times a second to carry out the intermediate steps needed to derive that evaluation. Perhaps such similarities between real human biology and effeciently engineered networks perpetuate AI's hope that truly intelligent machines are just around the corner.

Even the best device needs minor adjustments, called "tweaks" in engineering jargon. Tweaking the parameters one way will make the system better at some things and poorer at others. It will still be a functioning network, but a different one depending on how you tweak the parameters. Humans also have parameters that are not identical, a quasi-technical way of saying that we possess inborn differences for tolerating cognitive dissonance, such as paying more attention to either emotional or cognitive events. For example, one spouse hates dirt but doesn't mind disorder; the other is oblivious to crumbs on the floor but cannot stand papers or clothes strewn about.

Far more interesting human tweaks are seen in correlations between immune disorders and learning disabilities, or between immune disorders, high mathematical ability, and left-handedness in males. The top five percent of U.S. merit scholars, for example, are predominantly left-handed males with immune disorders. AI would maintain that such diversity is a technical issue, saying that we do not "yet" know how to account for such differences in formal language and therefore do not really "understand" them.

What AI believers do not understand is that the burden of humanists to show why qualia are necessary for intelligence has been satisfied. Now it appears to be their burden to replicate emotion before aiming for an intelligent machine that will succeed at thinking in general.

Chapter 5

DIFFERENT KINDS
OF KNOWLEDGE

The previous essay explored the technical limits of AI's efforts to dis-
embody the mind from the physical brain. The moral argument why
this is impossible rests on the fact of our having physical bodies as
well as a mind capable of different kinds of human knowledge.
Behind both stands the role of emotion in that subjective mental
state called consciousness.

I realize that science has done much for humanity and that we
largely owe to it the state of the world today, good as well as ill.
While often munificent, science can also be addictive and corrosive.
Because people hold exaggerated expectations of technologies they
can only superficially comprehend, science has replaced other means
by which individuals could make judgments with its own narrow
standards. Eons before modern science spawned the pervasive trust in
objective certainty that now dominates our thinking, humanity was
guided by other kinds of knowledge such as moral, æsthetic, and
judicial values that outlined one's relationship to nature and to
fellow humans.

Now, as MIT professor of computer science Joseph Weizenbaum
ruminates, "science has become the sole legitimate form of understanding." Alternate values have been cast aside. Attributing absolute certainty to the scientific method has "delegitimatized all other
ways of understanding. Once people viewed the arts, especially
literature, as sources of intellectual nourishment and understanding,
but today they are largely perceived as entertainments."[10] What

people insist on today is knowledge that has been "scientifically validated." Such a mind set is unable to admit and reconcile incommensurable values, a fault all too evident in the conflicts now tearing this planet's inhabitants.

Perhaps the most characteristic consequence of modern science is our rejection of direct experience that began with the appearance of the clock. The regularity of nature was replaced by an abstract string of moments whose passing was represented by a mechanism of gears, indicators, and numbers. Since then, what exists is being replaced by a representation of numbers with increasing rapidity and scope. What can be known has been reduced to what can be measured and calculated. Historian Lewis Mumford believed that the dissociation of time from nature and human events "helped create the belief in an independent world of mathematically measurable sequences: the special world of science."[11]

The clock seems to be the singular exception to the rule that machines are prosthetic. We come to view tools as extensions of our own body and senses, whether we refer to toothbrushes, typewriters, or space probes exploring the outer planets. Everyone has experienced how "that thing" the dentist has done becomes shortly transformed into "my teeth." Tools likewise come to feel a part of us quickly. Is it any surprise that the computer, which superficially mimics our intellectual abilities, should raise the view of the individual as a machine to a new level of possibility? The computer as a prosthetic extension of the human mind intimates immortality. If the egos enchanted with AI see it as a god-like creation of life in their own image, then is not the true AI believer this generation's Frankenstein?[12]

When *Homo sapiens*, the wise, became *Homo faber*, the builder, we began to transcend nature's limitations on our bodies. Tools transform the natural world and our perception of reality. Despite the enlargement of individual potential by tools, early humans still lived in accord with nature. Only since the scientific revolution, and mostly since the industrial one, have we completely conquered nature by our capacity to destroy life on earth. Those who say that computers are "merely a tool" imply that they are not important because tools themselves are not important in any fundamental sense. This idea is much mistaken because prosthetic tools inevitably transform both our own identity and our view of nature, and thus our psychological grasp of reality.

The pervasiveness of the computer metaphor has turned every problem into a technical one for which the methods of analysis are judged appropriate. Number crunching reduces us to objects. The solving of the genetic code and now the Human Genome Project, which will methodically catalogue our DNA sequences into an international database, solidifies the vision of the individual as an object. Taken with "breakthroughs" in genetic diseases, it implies that people may be conveniently altered or even designed to specification. Worst among its perceived accomplishments is a premature closure of ideas—the implication that everything that needs to be known will be known through it. Gene cataloguing is impressive, but we are more than our DNA. *All too often a scientific abstraction of reality soon comes to be seen as the total picture.* We are stranded with a single type of knowledge.

Logical formalisms illustrate the limits of objective knowledge. AI argues that we really never "understand" something until we can break it down into formal logic statements, a set of rules that state in precise and unambiguous language what to do from one moment to the next. But logical formalism is a terribly weak means of knowing. An experience is more than the sum of its moments. A recipe is not a meal, nor is a road map a journey. We might understand something expertly but be unable to formalize our understanding, particularly knowledge of procedures. Winning at poker involves more than knowing the rules, for example, and piano playing is much more than hitting all the right notes.

Except for a small set of formal problems that mostly concern physicists and mathematicians, the aggregate of human intelligence is directed to situations arising from unique biological and emotional needs. Machines will never fathom our deep knowledge because it is often unfathomable to ourselves. This is why we are so often at odds with ourselves and why attempts to "think through" solutions often lead to unhappy results. What we want, what we feel, and what we know are entirely different things. This human proviso is beyond the power of science to put into an equation.

Instead of formal language, humans speak in what is called natural language, real ones such as Hungarian or English. Natural languages certainly lack the precision and absoluteness of formal logic, yet they have served humanity well and machines have failed to duplicate them. Language theorists tell us that there are principles of linguistic understanding that cannot be formalized. Knowledge of the real world is gained through all of our senses as well as the experience of having a physical body. Our knowledge of the real world has a context—what we do with objects, what juxtapositions between elements are permissible, and a host of other factors that are exponentially large. How should computers deal with knowledge that is not linguistic? In computer vision, for example, the machine must "understand" what its samplings of light energy mean. Understanding goes beyond the recognition of patterns or even whole objects to context, which is a holistic understanding. Every culture has a set of unwritten rules about what is proper and how things are done. These expectations are learned by living in that culture. Even adults who have lived for years in a foreign culture are said to lack its interpersonal nuances.

There are additional ways of understanding that are hard to describe. There is knowledge that no machine can ever "understand" because it relates to objectives that are inappropriate to machines. No amount of logic nor programming bravado could capture what is learned from having a body and living in a culture. I cannot accept the idea that machines can do my thinking, because the basis on which a machine must base any decision is necessarily inhuman.

Chapter 6

▲ ■ ●

THE EXPERIENCE
OF METAPHOR

The coherence of metaphors and the beliefs that arise from them are not based in logic or rational thinking. They are rooted in concrete experience, which is what gives metaphors their meaning.[13] Linguists, philosophers, and objectivists will laugh at this premise because metaphor is traditionally considered an abstract rhetorical or poetic device. But I want to present the case that everyday language and actions are permeated with metaphoric concepts that are based in physical experience.[14]

I mean that metaphor is experiential and visceral, an irrational transfer of connotations from one thing to another. The emotional, irrational self is wise beyond knowledge, and we can see this wisdom in the way metaphor physically encapsulates our relations with the world. While metaphor is a means of seeing the similar in the dissimilar, it is emphatically not rational analysis.

A system of mental concepts determines how we think and act. We are not normally aware of our conceptual system; we just think and act along certain lines. Concepts structure what we perceive and how we relate to other people, thus centrally defining everyday reality. If our conceptual system is metaphorically based, then the way we think, what we experience, and what we do must also be metaphoric.

The view that metaphor is merely language perpetuates the view that the world is dispassionately objective, meaning free from human concepts of it. However, concepts are not defined by fixed properties but in terms of how we interact with objects. In other words, understanding grows out of the entire scope of our experience.

The objective person claims to comprehend something in terms of its inherent properties, some disembodied ideal much like Aristotle's common sensibles. Let me show this to be false through the most subjective example I can think of, namely LOVE. Dictionary writers defining love allude to affection, sexual allure, and the like. Metaphorical comprehension sees love as a JOURNEY, MADNESS, or a BATTLE—things grasped in the course of experiencing it directly. For example[15]

LOVE IS A JOURNEY

Look at *how far we've come* only to *go our separate ways*. It's been a *long, bumpy road* and this relationship *isn't going anywhere*. It's *on the rocks*.

LOVE IS MADNESS

I'm *crazy* about you and *insanely* jealous. You drive me *wild* and make me go *out of my mind*.

LOVE IS A BATTLE

She is *besieged* by suitors who *pursue* her *relentlessly*, causing her to *flee* their *advances* and *fend them off*. The *tactics* they use in *fighting* over her are unbelievable.

▲ ■ ●

Trying to write down an objective definition of love reveals the concept to be almost entirely metaphoric. A metaphor is often defined as experiencing one thing in terms of another, as the metaphorical knowledge of love illustrates. Metaphoric understanding is the ability to perceive similarity among seemingly dissimilar objects. As Aristotle put it, "It is from metaphor that we can best get hold of something fresh."[16]

The easiest metaphors to understand are those based on simple spatial directions such as *up*. We change our physical orientation during activities such as standing, sleeping, climbing, or driving. Since a physical orientation is central to having a body, orientation is central to our conceptual system. That is, the structure of our spatial concepts emerges from our direct physical experience.

CONSCIOUS IS UP; UNCONSCIOUS IS DOWN

Wake *up*. I'm *up* already. I'm an early *riser*. I *dropped* off and *fell* asleep. The patient *went under* anesthesia, *sank* into a coma, then *dropped* dead.

CONTROLLING IS UP; BEING CONTROLLED IS DOWN

He's *on top* of the situation, in *high* command, and at the *height* of power in having so many people *under* him. His influence started to *decline*, until he *fell* from power and *landed* as *low man* on the totem pole, back at the *bottom* of the heap.

GOOD IS UP; BAD IS DOWN

High-quality work made this a *peak* year and put us *over the top*. Things were looking *up* when the market *bottomed out* and hit an all-time *low*. It's been *downhill* ever since.

RATIONAL IS UP; EMOTIONAL IS DOWN

I *pulled myself up* from this sorry state and had a *high-level* intellectual discussion with my therapist, a *high-minded*, *lofty* individual. My heart *sank* and I was in the *depth* of despair, unable to *rise* above my emotions.

▲ ■ ●

The physical bases for these metaphors is that most mammals sleep lying down and stand up when awake. Well-being, control, and things characterized as good are all *up*. Since we control our physical environment, animals, and sometimes even other people, and since our ability to reason is what gives us this control, *control is up* implies *human is up* and therefore *rational is up*.

Cultural values unavoidably color mental concepts. For example, thinkers in the Age of Enlightenment asserted that "men are lofty creatures" by virtue of their advanced power of reason. Perhaps one factor leading to the view that reason was supreme was an attempt to separate us from uncultured barbarians and the undesirable traits of "lower" animals. So, metaphors like *rational is up* and *emotional is down* have both physical and cultural biases.

Spatial orientations like up-down, front-back, and center-periphery are the most common ones in our system of concepts, but given the variety of ways we interact with the world, there are others. We make inside-out distinctions between reason and emotion, for example, and generally characterize rationality as up, light, and active, while the emotions are down, deep, and murky—passive, irrational passions over which we have little control. Intellectual functions of the brain are called "higher" while the emotions and habits are "low."

Anthropologists tell us that the major orientations of up-down, in-out, central-peripheral, and active-passive exist in all cultures. But which concepts are most valued varies from place to place. Some cultures prize balance, whereas America seems taken with the extremes of *up* or *down* orientations.

We see that forming metaphoric concepts is like culling tidbits of our experience and then treating them like autonomous entities we can rearrange. Our interactions in space yield orientational metaphors. Other experiences give rise to what are called ontological metaphors, ways of treating events, actions, emotions, and ideas as self-contained objects. The influence of culture is what elaborates ontological metaphors. We can elaborate *the mind is an entity* into *the mind is a machine* or *the mind is a brittle object* and get the following:

THE MIND IS A MACHINE

We are *cranking out* a lot of paperwork. You could see his *wheels turning*. Their proposal just *ran out of steam*.

Compare this with the results of a different elaboration, namely,

THE MIND IS A BRITTLE OBJECT

He *cracked* under pressure. It was a *shattering* experience. You *bruised* his ego.

Metaphors emphasize some facets of an object but hide others. The *machine* metaphor paints the mind as having a source of power, an expected level of efficiency, an optimum production capacity, and an on-off state; but it hides the vagaries of thought, its ability to deal with fragmentary information, and other abilities resulting from its subjective quality.

By switching metaphors, we alter how we comprehend something and thus alter reality. Words cannot change reality but changing our concepts does alter what we perceive and how we act on those perceptions. Ontological metaphors are so pervasive that they seem natural and self-evident descriptions of mental thought. It never dawns on us that they are metaphors. Ponder the experience implicit in the following.

UNDERSTANDING IS SEEING; IDEAS ARE LIGHT

I *see* what you are saying. It was a *brilliant* remark and a *clear* discussion. Your point of *view* gave me the *whole picture*. Their proposal is *murky*, the ideas *opaque*, and their premise is *transparent*.

EMOTION IS PHYSICAL CONTACT

The verdict *bowled him over*. I was *struck* by his generosity. His donation *made an impression* on me. That model is a *knockout*. I was *touched* by their kindness.

You can see how different metaphors produce different flavors of a given concept. The intuitive appeal of a concept rests on how well its metaphors fit our actual experience. One factor contributing to the irrationality of the human mind is the conflict among metaphors that arises from *real differences in their physical foundations*.[17]

For example, "That's *up in the air*" and "The matter is *settled*" are each physically consistent with "I *grasp* your meaning." If you can grasp something, you can examine and understand it, and things are easier to grasp if they are down rather than flying up in the air. Thus, *unknown is up* and *known is down* are coherent with *understanding is grasping*. However, *unknown is up* is inconsistent with the orientational metaphors *good is up* or *finished is up* (e.g., "Finish up this last piece of pie").

Logic demands that *finished* be yoked with *known*, and *unfinished* with *unknown*. But our experience disagrees. We do not consider the unknown to be good, and the physical experience leading to *unknown is up* is furthermore totally different from that on which the two incongruent metaphors are based. This shows how the ability to be at odds with ourselves or the ability to simultaneously hold conflicting beliefs is based once more not on reason but on physical experience.

Chapter 7

▲ ■ ●

EMOTION HAS A LOGIC
OF ITS OWN

Plato said that we are prisoners of our feelings and that we should therefore hold fast to the sacred cord of reason lest we be lost. Euripides declared that folly occurs only when desire conflicts with reason.

Aristotle, on the other hand, argued with Plato that emotions have a logic of their own and must be understood on their own terms. He asserted that they were not simple animal passions unleashed, but were a complex part of our thinking.

There is an interesting schism in the way contemporary philosophers and the general public view emotion. Philosophers agree that emotions are complex states involving valence, salience, desire, judgment, and action. That is, the nature of emotions is partly rational.[18] The public, however, is more inclined to favor Plato's dichotomy that emotions conflict with and threaten reason.

For example, in *The March of Folly*, historian Barbara Tuchman argues that from Troy to Vietnam, leaders have pursued policies contradictory to their self-interest, even after the negative repercussions have become plain. These decisions were short-term judgments that had an emotional bias and that history judged to be follies, grand miscalculations whose consequences were predicted at the time by minority voices. In asserting that emotion, not well-reasoned statesmanship, steers the ship of state, and that "the rejection of reason is the prime characteristic of folly," Tuchman illustrates the widespread popular belief in the Platonic dichotomy.

The Platonic view perceives a strong dichotomy between reason and emotion, while the Aristotelian view sees a much weaker dichotomy. Dr. Ommaya nicely eliminated the dichotomy altogether by proposing that consciousness is a type of emotion. Emotion and reason are interdependent because their anatomy is interdependent, but aspects of them can be separately apprehended. Logical reasoning makes us feel that "we" handle the process. The logic of emotion, however, is beyond our control. Therefore, our energy is better spent trying to understand its conclusions rather than change them by throwing reasons at it.

I have likened reason and awareness to the tip of an iceberg. Being conscious of our thoughts seems to have emerged with the appearance of brain areas capable of the symbolic precursors to language. But language is hardly the only means of self-expression. The brain can direct the hands and body in piano playing, painting, mime, dance, or other creative acts. These non-linguistic motor outputs express a highly sophisticated self-awareness that is strongly allied with the æsthetic capacity that I have argued depends on emotion rather than reasoning skill. My point is that when we experience the light bulb going off, we first discern a feeling of recognition and coherence, followed by a conscious recognition, "this is it."

That the conscious awareness of an insightful recognition is secondary to the recognition itself is splendidly shown in the neurological condition of prosopagnosia, meaning "not knowing faces." Patients are no longer able to compute the details of a particular face from the general class of faces and attach it to a memory of who that person is. This is true even if that person is a spouse or someone the patient has known for years. One can show, however, that another facet of their mind does recognize the person to whom the face belongs. Galvanic skin resistance (GSR) reflects sympathetic nervous output related to emotional structures that are crucial to memory, including the memory of faces.

When patients with prosopagnosia are shown a picture of someone they knew well before their illness, two contradictory things occur. The cognitive mind says that it does not know who that person is, while a sharp GSR response betrays that recognition has indeed taken place at an unconscious level. In other words, recognition can be dissociated from conscious awareness of it. That the logic of emotion is dissociated from reason is most readily evident when it operates in creative and spiritual realms. For example, we may be struggling with a vague idea, just going with a feeling that directs us toward an inarticulate goal. Often we say, "I'll know it when I see it," or "I'll know it when I've done it." The subjective part of the emotional brain is attuned to a deep source of wisdom and engaged in a process to which the cognitive mind has no access. Only after the logic of emotion creates order does its coherence become visible to the cognitive mind, which is then free to "explain" the solution. The author Flannery O'Connor put it perfectly when she said, "I write because I don't know what I think until I read what I say."

I defined insight as fathoming a relation between different premises, a capacity that depended on the emotional organization of our brain. A solution toward which we struggle does not always become apparent by attacking problems with the mind. Solving a koan, a mental puzzle given to Zen students, illustrates this truth. To work on a koan the student must be eager to solve it and face it without thinking about it. The more one attacks a koan with the intellect, the more impossible a solution becomes. "Two hands brought together make a sound. What is the sound of one hand clapping?" This is a famous koan. If you think there is no such sound you are mistaken. A Zen koan is utter nonsense to outsiders, but a gate to enlightenment to Zen students.

Things like koans deliberately force us beyond analysis and heighten our appreciation that the cognitive and emotional minds are but two aspects of the same mind. Sometimes I visualize the mind as a gemstone, a single entity with many facets, to help remind me that the cognitive mind is but one of several mental entities. The idea that we have more than one mind was first suggested by A. L. Wigan in 1844. In *The Duality of Mind*, he describes the autopsy of someone he knew well in whom one cerebral hemisphere was totally absent![19] Wigan had the sense to see that a single hemisphere was sufficient to be a person. He suggested that the brain is not a single organ of two halves but a closely apposed pair, just as the kidneys or lungs are paired organs. Wigan concluded that if one hemisphere was sufficient to have a mind, then the customary two makes having two minds inevitable, and however synchronous they might usually be, there must be times when they are discrepant. Wigan's observation suggested a physical basis for that internal conflict so characteristic of humans.

Wigan's magnificent speculation raised little stir at the time. More than a century later, the split-brain operation dramatically showed that having two hemispheres means that we possess two minds that differ in content, mode of organization, and even in goals. What further enterprise in neuropsychology has made clear is that we have multiple modes of thinking, not just two, and that most of them occur at a level about which we have no sense of pulling the strings. The general subject of unconscious knowledge is called *subception*, meaning "below awareness," to contrast with *perception* (Latin *percipere* = to receive), which means to take into awareness. Blindsight, prosopagnosia, and the ability to learn during anesthesia are some of the better-known examples.[20]

The possession of multiple minds may illuminate the phenomenon of projection, the tendency to cast our own feelings, desires, and fears onto others. If so, then a sense of certitude, the feeling of a presence, or some other exalted state of consciousness may be just a seldom-seen aspect of our own selves projected onto the environment. On the one hand, you might be disappointed to ponder that the experience of a profound insight, a clairvoyant vision, or a divine presence might have been "just us" all along. On the other hand, projection implies, in the context of creativity, that the muse resides within each of us and is not an external agent that visits only a chosen few.

What split-brain surgery did for the right-brain-left-brain dichotomy, multiplex anatomy will likely do as irrefutable proof that we possess multiple minds. The new challenge is to see our multiple minds as unified rather than fractured and warring with each other. For example, people often consider scientists and artists as alien breeds reared apart, as if imagination belonged to one and not the other, or as if fact and feeling were mutually exclusive. This opinion is widespread despite numerous examples, particularly from the Renaissance, showing that scientist, poet, painter, scholar, and philosopher have lived side by side in the same head without difficulty. I prefer to think that science is about how the heavens, the earth, and ourselves are constructed, while art is about the finished product.

In a way, the multiple facets of mind are the human equivalent of the duality principle in physics, which says that light is simultaneously a wave and a particle and that any experiment designed to demonstrate one property makes it impossible to observe its complementary one. Our metaphoric concepts, based in the logic of experience, do the same thing: in emphasizing one aspect of an object, they hide others. By analogy to the duality principle, our minds are simultaneously analytical and intuitive, appositional and propositional, holistic and sequential. Even though we scurry among these mental facets we cannot occupy more than one plane at a time, nor is any but the linguistic facet fully accessible to awareness and what we call reasoning. Yet the logic of emotion leads us to see that all facets are the expression of a single person.

Chapter 8

▲ ■ ●

OTHER PEOPLE'S EXPERIENCE

I love to write. It is an activity I have engaged in since I was a teenager. But for even longer I have listened, because I think there is nothing more interesting than hearing about the experience of other people. Storytelling is central to human experience and validates the importance of direct experience, not just to the one whose immediate experience it is but to others as well. Stories are all that remain of some of the greatest civilizations. Even cinema, which is only a hundred years old, is a visual translation of a much older oral one.

Storytelling is vitally important even though we are often quick to trivialize it as entertainment. When people are under psychologic stress or require answers to universal questions, a story comes to the rescue by sharing the experience of others. We want stories for succor, for assurance that we do not confront human problems alone and that someone has walked this path before us. Stories convey meaningful experiences and prompt us to action, allay our anxieties, or fulfill other psychic functions.

We long to know the experience of other people. To satisfy our psychic anxieties, we ask for stories, not logical propositions, rational explanations, or laundry lists of facts. In many circumstances, only hearing the experience of those who have gone before us can satisfy our psychic need. The centrality of stories in every culture and the psychic satisfaction they instill indicates how important it is for us to *attempt to reach the quality of another's subjective experience.* The long view of human history shows the vital importance of qualia.

Being enamored of writing, I am naturally interested in that carefully controlled flight of imagination called fiction. Good fiction totally absorbs the reader's mind, dissolving the real environment in which the reader holds the book. What impresses me most is how fiction accentuates human irrationality and how in fiction the astonishing and irrational is everywhere acceptable.

Good stories detail human extremes, people and situations that are worse than bad or better than good. Literature is not about what happened but what is plausible, even if impossible. Fables and science fiction are examples. Or consider Gregor Samsa turning into a cockroach in Kafka's short story "Metamorphosis." Readers don't put the book down after that first paragraph and say, "This is ridiculous." Its plausibility in how it is described is what makes it believable. It seems no one is much interested in reading about average people or ordinary events. But lying is never permissible because, as Aristotle pointed out, all fiction must be true no matter how improbable. This necessity curiously gives the writer a high moral obligation, but, more importantly, it means that well-crafted fiction leads to the discovery of universal truth, which readers experience as a satisfying emotional recognition. The poet can probably discover more truth than the historian can.

Satisfying art is a product of deep knowledge and understanding within the artist. It is true that art is informed by the intellect and with acquired technique. But the function of the artist is to penetrate the visible world to illuminate the mystery behind it. That mystery is a ground of universal truth that supports the human condition. If successful, the artist's expression resonates within the inner life of the reader, viewer, or listener who experiences what I have called an intuitive recognition. Ultimately, the art of fiction is not an intellectual achievement, but an emotional one in which intellect serves only to articulate the human truth, not to explain it.

Chapter 9

▲ ■ ●

THE DEPTH AT WHICH
WE REALLY LIVE

Have you ever noticed how popular culture repeatedly deals with romance, love's fulfillment, and finding your bliss? In stark contrast, intellectual and academic writings neglect the importance of obtaining your heart's desire or dismiss it altogether. Serious academic literature on love is tellingly absent, while popular venues such as films, videos, novels, and poetry often make it their focus. The vicissitude of romance is possibly the central tale voiced during the millions of annual psychotherapy visits. These examples illustrate the fundamental split between how we treat thinking and feeling.

Mentally, we split our existence into objective and subjective parts that respectively deal with the world's external demands and those inner concerns that are personally relevant. We try to make sense of the world by creating dichotomies and thinking in categories. Unfortunately, reality is not the same as the words used to describe it; we often fail to grasp what a burdensome imposition intellectual categories are, and how much they diminish our meaningful experience.

Since our decisions are governed more by feelings than by logic, it is no accident that popular culture rather than academic treatises take the pulse of what we long for. Popular culture thrives on and celebrates the subjective needs of the human psyche, speaking directly to us and resonating to our depths. We "get it" without having to have anything explained. A movie, a myth, a novel, or a painting that has to be explained has failed as a symbol.

Emotional affirmation strikes a chord precisely because it does speak to the heart instead of the brain. This is perhaps why the film-maker, artist, writer, and even the psychic are so popular in our culture. They are the cultural champions of the individual's emotional needs. They not only concede but celebrate that realm of subjectivity that society rejects as weak, too feminine, or unrealistic. They confirm that the fulfillment of our longings and the desire for inner expression is not only meaningful but vitally important.

A pervasive distrust of our irrational intuitions and emotions is evident in stock phrases like "Sorry, I wasn't thinking." I have never heard anyone say, "Sorry, I wasn't feeling." We are prone to identify ourselves with the rational, the external, and the objective. This is particularly evident in psychiatry, whose central activity is called psycho-*analysis*. How strange that we identify so strongly with the rational mind while we connect most deeply with life through our emotional psyche. In response to this curious split, I propose that it is more satisfying to *evoke* the psyche rather than to *analyze* it.

Evoking the psyche rather than analyzing it moves you away from the surface of external concerns and explanations so that you can grasp the depth at which you really live. It seems obvious that you should understand yourself better than you could ever understand anyone else. But when you actually try to fathom what you do, what you feel, and what you believe, then such efforts to understand take you beyond yourself. This is the definition of transcendence.

Ineffable, noëtic, and *transcendent* are words that point to something behind the surface, behind what the philosopher Kant called "ordinary experience as we know we have it." William James said that the *ineffable* defied expression. "Its quality must be directly experienced, it cannot be imparted or transferred to others." *Noëtic* refers to knowledge which is directly imparted, an illumination that is accompanied by certainty. *Transcend* means "to climb over or beyond." It refers to that which we cannot name. These three words all point to the existence of inner knowledge and a dimension beyond which words cannot reach.

All three views betray an underlying hunger for understanding. Objectively, you seek to understand a world of external objects and relationships; subjectively, you seek an internal understanding that makes life worth living. The pure objectivist believes in a world made of objects with inherent properties about which one can utter statements that are absolutely true or false. The pure objectivist feels secure in believing that science has a method to avoid the subjective limitations such as errors and bias that make the human mind objectively unreliable. The pure subjectivist, on the other hand, rejects the impersonal and abstract, perhaps turning to the Romanticism that originally grew to counteract the ascendancy of the scientific method. Turning to art and nature was the means by which Romantics hoped to recover their humanity that had been lost in the industrial age.[21]

Both the scientific objectivist and the romantic subjectivist view the individual as autonomous, and both try to overcome the individual's existential alienation and separation from nature. The scientist tries to rejoin nature by conquering it; the Romantic communes with nature or becomes absorbed into it. A third view based in experience stresses interaction: we cannot live in the world without changing it or being changed by it. The meanings of metaphor, for example, are grounded in physical experiences that structure the conceptual system of what we believe and how we act. These actions in turn alter the world.

Artist-scientists are often assumed to live compartmentalized lives, one realm in which they act scientifically and another in which they can be creative. In other words, the objective and subjective views define themselves in contrary terms and seem to exist in separate domains. How can the third view based on experience get past this dichotomy?

Because metaphor joins reason and imagination, the conceptual system on which reality is based is in part imaginative. Likewise, creative ideas are partly rational in nature. Objectivity fails to see that our system of concepts is metaphoric, involving an imaginative understanding of one thing in terms of another. Subjectivity fails to see that even our most imaginative flights occur in the context of objective experience gained by living in a physical and cultural world. The elaboration of metaphors, for example, is an imaginative form of rational thinking, yet romanticism denies that human thought is constrained by any context.

The experience of living cannot be neatly carved up into wholly objective or subjective portions. Fortunately, the middle view based in experience does not need to be absolute. It produces a truth relative to our system of concepts, a system rooted in and constantly refreshed by experience. Neither the objective nor subjective views alone completely fathom that we understand the world by living in it. Experience is noëtic.

We are grasping for a sense of unity because modern life does not fulfill the needs of the human spirit. By embracing subjectivity, the Romantic tradition carved out a niche for itself in the realms of art and religion. In terms of real power, however, modern life is driven by technology, politics, and economics, surface issues that worry the rational mind. Precisely because these superficial drives are so strong, we habitually ignore the depth at which we really live.

It is a curious fact of modern life that we live on the surface and deny the force and reality of our inner experiences. We have been brought up with the shoulds and oughts, the voice of the editor that tells us what we cannot or must not do. We scold ourselves with this inner voice: "if only I had," or "if only I had not." We willingly deny our deepest desires in order to live as society tells us to.

What happens is that we often end up living our parents' lives, or striving for those things that others think are so important, such as money, position, and so-called power. Success, for example, is a category that we impose on our experience. We treat an idea of success as a real thing, and say we possess success upon acquiring the external conditions that define it. The difference between external, categorical success and that which satisfies our internal needs is enormous. But following the dictates of our inner needs always produces greater satisfaction. To follow the dictates of society, even though it leads to a type of prescribed happiness (that which society considers good), is to live an inauthentic life. In a most astounding way, we have been trained to set aside or postpone our feelings, our dreams, our desires, and the whole compartment of our subjective needs.

The first step in breaking through to the transcendent is putting aside the idea that we have to choose between objective and subjective views of reality. Many aspects of human experience cannot be conveyed by objective facts, nor is there any escape from subjectivity. In addition to a detached, objective view based on externals and a subjective view based on our inner life, there is a third choice grounded in experience, through which noëtic understanding is found. This is the depth at which we really live.

Chapter 10

REASON IS THE ENDLESS PAPERWORK OF THE MIND

Two facets of our multiple minds that I wish to single out are the cognitive mind and our emotional one. The cognitive mind is the one that speaks out loud to others and silently to ourselves in the form of internal dialogues. The cognitive mind must have reasons; the emotional mind is attracted to experience.

The cognitive mind is always future oriented and very much concerned with desire, possession, and control. If we get what we want, the cognitive mind is momentarily relieved not because it has the object of its desire but because desire itself is temporarily quenched. Desire, and not the object, is what seems attractive.

The cognitive mind is interested in analyzing and explaining because it thinks it can control what it understands. If it has reasons, then it believes it can exert its will to change circumstances to suit its desires. It never occurs to the cognitive mind to change its desires to suit circumstances, which is what the emotional mind would do.

We often refuse to accept a situation contrary to our desires because our cognitive mind identifies so strongly with our desire for it to be different. It must always be doing something and never accepts what is already done and simply there. The most difficult thing for the cognitive mind to grasp is that sometimes there is nothing for us to do. The emotional mind, on the other hand, lets objects and situations present themselves as they are. Eventually we learn acceptance. When we find ourselves in contrary situations, self-analysis rarely helps because all we end up with is the mind's conclusions, which again are reasons that prime the pump for the cognitive mind to exert its will even further and try to change what it cannot change.

While the cognitive mind is preoccupied with doing something and looking to the future, the essence of the emotional mind is attending to the present and being quietly receptive. When we are in touch with the heart, we can feel even the strongest of emotions without creating confusing mind dramas. That is, we can feel our feelings without getting lost in them, without jumping to conclusions and judging ourselves negatively by virtue of the many possible reasons that the cognitive mind can whip up.

Granted, it is not easy to live in the present, even though it is where we always are. Given that we have no choice, is it not fascinating how much energy we spend reaching back to relive the past or longing for a rosier future? The technological rush of our society also tries to sweep us fast forward, but through practice we can anchor ourselves in the present more firmly and learn to accept things as they are. By pushing, insisting, and forcing, the cognitive mind usually creates the very opposite of what we want; by accepting, opening up, and being patient we invite what we want to come to us. The emotional mind teaches us to respond to situations differently from our usual way.

The future-oriented cognitive mind always thinks that things will be better later—when we are older, more experienced, smarter, better looking, financially more secure, or ten pounds thinner. If we cannot accept ourselves and circumstances as they are, then we confuse ourselves with resentment and guilt. Inner growth is impossible in such a negative climate. The emotional mind says, "I'm OK as I am." We enjoy what we are doing right now. We accept things as they are right now. When we are flexible and open to experience we enter life more fully.

This distinction between the cognitive and emotional minds has everything to do with transformation. We are always becoming. But resistance and strong will impede our self-evolution. We refuse to trust what we cannot see with our own eyes, so we either resist change or try to force it to happen now. We tear open the cocoon to see the butterfly, pry open the rosebud to smell the blossom, and ruin both so that there is nothing left to savor. A shift from the cognitive mind to the heart shows that these efforts are misguided. We must therefore practice letting go, which means doing nothing. "Do nothing, and everything will be done." The wisdom of this proverb says that contrary to appearances, things are constantly changing. We have to trust what we can't see.

We may feel, when situations are contrary to our desire, that unless we do something we will go out of our minds—which is exactly where we do need to go. Our body functions without our having to do anything. Our body knows how to digest the food we eat and extract the oxygen in the air we breathe. Our organs maintain the balance of our internal chemical ocean. This and much else is orchestrated without our having to think about controlling it. Such is the effectiveness of letting go and trusting.

As Pascal said in *Pensées*, "The heart has its reasons, which reason knows not." Reasons are nothing more than the endless paperwork of the cognitive mind which always hunts for explanations so that it can enforce its conclusions.[22] If we want to enter the emotional mind, we have to let go of our belief in reasons, including such beliefs that we are not creative, nobody loves us, we will never get ahead, that we cannot express who we really are, or that all there is is what we see on the surface. Because we never know what the next moment will bring, transformation could occur at any time. This is a very encouraging thought.

The only action we have to take is to decide whether we are going to stay closed or open up, accept experience as it is or rationalize it to death. This is what Joseph Campbell meant when he said that what we seek is not the meaning of life but the "experience of being alive."

Chapter 11

▲ ■ ●

SCIENCE AND SPIRITUALITY

The words "like us" seem to dominate discussions of other brains, other species, and artificial intelligence. Embedded in this phrase is the assumption that "like us" is as good as it gets. However, as our consideration of the anthropic principle showed, many humans have adopted an I-Thou attitude that accepts other creatures as different without needing to judge them as inferior.

This attitude, or the lack of it, reflects our spiritual sensitivity. When scientists recognize themselves to be spiritual, believing in a deity or some force greater than themselves, then the attitude of not being "like us" does not really arise. Persons who claim to be non-religious often act as their own gods, in the sense of believing that they possess all the powers for creating whatever worlds they please with the tools of technology. They believe they can fully turn nature to human service. The distinction between the spiritual person and the non-spiritual one who likes to be called "objective" boils down to what kinds of values each is willing to employ.

The classical measure is yourself. That is, you judge others in terms of what you know about yourself. Both spiritual and objective persons adhere to standards based entirely on faith. The belief of a spiritual person in a larger force can only be accepted on faith. The atheist's conviction that no realm exists beyond ordinary experience is unprovable and so also a matter of faith. It is a catch-22 whereby an ardent atheist or strict objectivist cannot really claim to be rational.

To be consistent, both the ardent atheist and the strict objectivist, those who believe that all mental states can be reduced to physical states of the brain, must end up like Roger Sperry or Jacques Monod, both Nobel Prize-winning atheistic scientists who wish to create a value system based on science. It is not difficult to create such a value system, but it is one I would fear.

Alasdair MacIntyre, the moral philosopher who wrote *Whose Truth, Whose Virtue,* cautioned that those who espouse "objective" viewpoints always think that their own arguments are the most rational, logical, and convincing. "My civilization, my culture, my method, and my values are better than yours," they say.

The behaviorist B. F. Skinner offers an example of extreme faith in objectivity. "In every walk of life," he says, "from international affairs to the care of a baby, we shall continue to be inept until a scientific analysis clarifies the advantages of a more effective technology. In the behavioristic view, man can now control his own destiny because he knows what must be done and how to do it."[23] What this really means is that B. F. Skinner "knows what must be done and how to do it." Like everyone who believes in objectivity, he is sure that he knows what is best for the rest of us.

The flaw in worshipping objectivity is this: it is possible to have an objective view of anything, but only from your own subjective point of view. You cannot have a subjective evaluation of a species other than yourself, for instance. Hence, you cannot know what it is like to be a bat, a whale, or anything other than yourself.[24] Every subjective experience is connected with a single point of view. The error of non-spiritual persons who place reason and objectivity above all else (as their gods, you might say) is in trying to develop an objective view from nowhere, a view detached from other values. Perhaps we can imagine a view sitting out in space, but the more we think about it, the more we see that it is impossible to have a view from nowhere without beginning with a view from somewhere. That somewhere is yourself. It is difficult to imagine what the *objective* character of an experience would be like. "After all," asks philosopher Thomas Nagel, "what would be left of what it was like to be a bat if one removed the viewpoint of the bat?"

A crucial principle that I hope has become clear is that objectivity is impossible without prior subjective experience on which to build. If the subjective character of any experience is comprehensible only from one point of view, then any shift to objectivity—that is, a detachment from your own specific viewpoint—*only takes you further away* from the quality of that experience.

We live in an uncertain world, and the grasp for certainty is a driving force of both science and religion. Religion and mythology sought to make sense out of nature's unpredictable power, while the quest for control is inherent in all technology. We seem to have two incompatible systems in collision, with most folks having opted for faith in science long ago, ditching inner knowing along the way. It is a mistake to think that one must make an either-or choice. We need to live with both and need both to live. We cannot discard our objectifying impulse, but we must insist that it learn to live alongside the inner perspectives that can neither be denied nor objectified.

It is a truism that science tells us how the physical universe "really is," but its rejection of direct experience leads to a distorted view of it. Science simplifies reality by leaving out whatever fails to fit the conceptual framework with which it is working at the moment. Aldous Huxley expressed this nicely back in 1946:

> The scientific picture of the world is inadequate for the simple reason that science deals only with certain aspects of experience in certain contexts. All this is quite clearly understood by the more philosophically minded men of science. But most others tend to accept the world picture implicit in the theories of science as a complete and exhaustive account of reality.[25]

Here is the flaw of the objective expert who "knows what must be done and how to do it." Factual concepts of life and human behavior are far less than a complete picture. Objective frameworks promote themselves as without values while simultaneously asserting an authoritarianism based on expertise, a value judgment in itself. Proposals for value-neutral, "objective" decisions never acknowledge that value choices have already been made. We must beware of abstractions that claim to capture the whole picture: their visions are detached from authentic human experience and do not even permit questions of values and subjectivity to arise.

The world is many things, and no single framework can hold it all, neither science nor art, neither analysis nor intuition. We need harmony between science and the spiritual, between the subjective and objective. Ironically, the problem is not so much science itself as it is the public's limited understanding of it, which too easily accepts knowledge produced by the scientific method as absolute and final.

Under ideal conditions, science itself is aware of its obligation to acknowledge what it does not yet and sometimes cannot ever know.

Not all who call themselves scientists worship at the altar of objectivity. One finds a continuum, with those who want to reduce everything to formal terms at one end, and scientists who keep company with humanists, artists, and spiritually minded people at the other. Einstein, in one of his more expansive moments, said, "The deeper we search the more we find there is to know, and as long as human life exists, I believe it will always be so." This is the message of religion too. Technology fosters an emphasis on external values, whereas the spiritual is about the inner life of anthropos and being able to touch the rapture of being alive.

The challenge is to go beyond the surface and contact this dimension which we do not fully understand but to which we wish to be open. Our own subjective life is the bridge to it. The proof of its existence is our own personal experience, whether spiritual rapture or the enlightenment (kensho) of formal meditation; insight gained from journal writing or other non-analytic exercises; being moved by an artistic or æsthetic experience; any creative act; feeling physical harmony in athletics, dance, or some other kinesthetic expression; or taking a journey of the mind in which the voyage is not an analysis of some problem but an intuitive leap and creative synthesis of ideas.

Our customary term to describe that which transcends what we ordinarily know is God. This turns many people off, either because they do not believe religion is relevant to their lives or because they distrust religion as an institution. Since "religion" has gained negative associations for many people, I prefer the term *spirituality*. The word *religion* comes from the Latin *religio*, often said to mean "linking back," implying a reaching back beyond material and temporal things, beyond pairs of opposites, and into the ineffable, transcendent state that is known by many names (nirvana, the state of grace, the divine presence, nature's realm, and so forth).

The external world is a world of opposites. Contradictions and dichotomies are necessary to the way that human minds ponder. When people talk about inner issues, the word God often comes up. God is an idea, a name whose reference is to something that transcends all categories of thought. Subjective experience implies the existence of facts beyond our ability to conceive. We can feel the existence of such facts without being able to objectify them, let alone comprehend them fully.

One of the features of faith is that it allows us to know both the inner world of the transcendent and the outer world of opposites. The deep nature of faith and mythology, the constructs that have built civilizations and informed human life through millennia, have to do with inner thresholds of passage, inner potentials of what humans are capable of experiencing and knowing. It has to do with the experience of life, while the mind has to do with meaning. What, for example, is the meaning of a flower? This is an absurd question because such things can only be experienced, not analyzed.

▲ ■ ●

Water is an ancient symbol of the unconscious. The waves of its ocean break on the shore of our awareness. We can stand on the shore and gaze into the water, but all we can discern is surface conditions. We can't fathom the depths from the shore. When we dive into the water we inhabit a different realm where our thought processes are different from those we used when standing on the shore. Here are the reveries from which creativity arises, the disjointed images and emotions of dream states, meditation, mysticism, and spiritual tranquility. It is in the substance of our unconscious ocean that we feel certainty, control, surety, a deep sense of knowing, not in the analytic chatter of the cognitive mind. When we pull ourselves out and stand again on the shore it is difficult, perhaps even impossible, to express what happened to us in that other realm or what knowledge we have gained. But we definitely know, we feel, that we have gained something.

Only a small part of inner knowledge reaches the surface of awareness. A further loss results from being able to express it only through the language hemisphere, which, as we saw, can sometimes be actively misleading. As Heinrich Zimmer put it, "The best things can't be said." The second best are those things we can only refer to, such as God, transcendence, and inner knowledge. The third best are the things our language hemisphere actually talks about.

As I see it, the artist's role is to see through the surface on which we usually live and into a transcendent reality, which is why, when we see a fortuitously composed work of art, our entire body responds with the "ah ha" of recognition. It is an æsthetic validation that cannot adequately be put into words.

This æsthetic sense of recognition is also the central validation of the spiritual person who says, "When I am in a religious state, it is the most beautiful thing there is." The certitude that "this is it" is all the validation necessary. I would point out that we also feel "this is it" during any insight or creative act, when things go well and our actions genuinely emerge from ourselves rather than from external pressures.

To be whole we have to explore both our inner and outer realities. We have both positive and negative experiences when swimming in the sea of our unconscious. And when standing on the shore of awareness, it is not just inspiration that washes up, but also the darkest truths about oneself.

Afterword

▲ ■ ●

CONTRADICTIONS AND CLUES: SYNESTHESIA AND CONVENTIONAL IDEAS IN BRAIN SCIENCE

Remember my colleagues' warning not to study synesthesia because it would ruin my career? "Too New Age," they cautioned. Well, I'm still here and many others have joined me.

For decades, most scientists who heard about synesthesia simply shrugged their shoulders or rolled their eyes. Some conceded its existence, but dismissed it as a mere curiosity or, worse, as just subjective imagination.

Since the publication in 1989 of the textbook *Synesthesia: A Union of the Senses*, the first English-language book devoted to the phenomenon, and *The Man Who Tasted Shapes* in 1993, there has been a renaissance of interest in synesthesia as neuroscience has relaxed its reflexive hostility to subjective experience. Today, researchers in some fifteen countries are studying it, and many doctoral candidates have chosen synesthesia for their theses.

Science has gradually become receptive once again to the existence of mental life, especially in the realms of memory, consciousness, and emotion. Following *The Man Who*, in which I emphasized the qualities of context and salience with respect to emotion, Damasio (1994) later emphasized the frontal lobes' role in evaluation—that is, in how intellect *informs* emotion as we make decisions and take action. Either too much or too little emotion results in poor intelligence. Taken together, our two viewpoints stress the pervasiveness and importance of emotion in human affairs. Many subsequent texts have extended and explored this theme (e.g., Moldoveneau & Nohria, 2002; Evans, 2003). *Psyche,* the on-line journal devoted to the multidisciplinary study of consciousness, began the symposium concerning synesthesia in 1995 (Cytowic, 1995), which interested readers may wish to retrieve or even contribute to.

If others have gradually come to accept the reality of synesthesia, they must now relinquish some received wisdom about how the brain works. Our concepts of how things work are but models, after all, reductions of reality that arise from human minds; history has shown repeatedly that reality has a way of making a mess of neat and tidy concepts. Like most exceptions of nature, synesthesia is forcing a paradigm shift. One cannot admit a wrecking ball and expect the house to remain standing. Paradoxically, however, the very thing that destroys simultaneously illuminates, and what remains may surprise us.

As inquiry into synesthesia has progressed from description, to nosology, to psychophysics and genetic analysis, the quality of tools for probing synesthetic brains has also improved stepwise from xenon-based rCBF, to PET, to fMRI. This progression now brings us to a new level of inquiry as we examine some of the challenges that synesthesia poses to current models of cognition. Four that I want to mention in this epilogue are blindsight (a category of subception), modularity, functionalism, and binding.

But first, I want to give readers a quick tour of synesthesia's general features as well as refinements in my thinking since the first publication of *The Man Who Tasted Shapes*. MIT Press and I together decided that to revise the text so as to incorporate new knowledge and answer the numerous questions readers have posed about synesthesia in general would be to ruin the story of Michael Watson. Instead, we use this postscript to flesh out details of synesthesia as a general phenomenon. As it turns out, Michael Watson's taste-touch synesthesia is quite rare compared to other types (table 3).

HOW COMMON?

As children, synesthetes are surprised to discover that others are not like them. Often ridiculed and disbelieved, they keep their atypical perceptions private. Nonetheless, the phenomenon remains involuntary and consistent throughout their lives.

Synesthesia turns out to be much more common than the estimate based on my original sample of forty-two individuals. Every population

Table 3. Types of synesthesia (n=365)	
Colored graphemes	66.8%
Colored time units	19.2%
Colored musical sounds	14.5%
Colored general sounds	12.1%
Colored phonemes	9.6%
Colored musical notes	10.4%
Colored personalities	4.4%
Colored tastes	6.3%
Colored pain	4.4%
Colored odors	5.8%
Colored temperature	2.2%
Colored touch	1.9%
Sound → touch	2.7%
Sound → taste	2.7%
Sound → smell	1.1%
Sound → temperature	0.5%
Taste → hearing	0.3%
Taste → touch	1.1%
Touch → taste	0.5%
Touch → smell	0.3%
Touch → hearing	0.5%
Vision → taste	1.9%
Vision → hearing	1.1%
Vision → smell	1.1%
Vision → touch	0.8%
Smell → sound	0.3%
Smell → touch	1.1%

About 40% of individuals have multiple synesthesias. From 365 cases complied by Sean A. Day, Ph.D., moderator of The Synesthesia List (available at http://www.users.muohio.edu/daysa/), with permission.

estimate performed so far has methodological shortcomings, and the least objectionable study suggests that some type of synesthetic experience occurs in at least 1 in 2,000 individuals. It could be more common, perhaps approaching 1 in 200 or even 1 in 20. It seems all that we agree on is that the first digit is a "2," and so we must await new canvassing to pin the number down.

Sensing color upon hearing, reading, or thinking of letters and integers—which I almost set aside as mild cases that were less interesting than those like Michael's or Victoria's—turns out to be the most common expression of the trait, accounting for two-thirds of instances. *Lexical synesthesia* is the term used to refer to a trigger of letters or words.

Most often, it is the written element, or *grapheme*, that evokes color, so homonyms look different because they are written differently. In about 10 percent of instances, however, it is the sound of language, or *phoneme*, that evokes the colored shapes that arise, move, alter, and decay—somewhat like fireworks. In the broader category of *colored hearing*, music and environmental sounds evoke photisms. For instance, hearing a doorbell might make one person see brown and gray triangles drifting off to the right as they fade; a dog's bark might produce shimmering circles moving out from the center for another; and the whoosh of a furnace ignition might produce a stack of colored lines for a third.

SUBJECTIVE REALITY OR MERE IMAGINATION?

Often throughout its history, synesthesia had been dismissed because the condition is revealed only through an individual's self-reported mental state. There is no test for it in the usual sense of that word. The complaint that introspection is inherently unreliable and therefore impermissible as scientific data has a long history. In the nineteenth century, psychophysicists such as Gustav Fechner tried to formulate laws regarding sensation and perception based on observers' reports, taking as a given that mental states exist. Scientists in the twentieth century, however, consistently strove to eliminate the subjective role of a human observer in gathering empirical data. Within psychology, the triumph of behaviorist theory further ensured that inquiry into mental life remained taboo for decades.

Because a technological focus dominated science in general and medicine in particular, my neurology colleagues unsurprisingly asked what

Michael's CAT scan showed. It was, after all, the machine marvel of its time in the late seventies. In questioning synesthesia's reality, they sought what any person claiming to be rational would have: a third-person technological verification of a first-person experience. Technical corroboration is one thing; but the sweeping assumption that anyone's personal experience is invalid is quite another. Even the current craze of functional brain imaging, which poses as anatomically objective, starts with what one wants to verify objectively: *the subject's state of mind*.

Synesthesia refuses to be ignored, affirming loudly that subjective mental worlds do exist. Among other things, therefore, synesthesia grants us an opportunity to examine the dichotomy of objective-subjective experience. But its importance goes deeper than that.

A LINK WITH THE PAST

Compared to the hostility of modern objectivists, a fair number of earlier scientists accepted synesthesia as a genuine phenomenon ever since Sir Francis Galton's 1880 report in *Nature* on "visualized numerals"—if only because the individual stories sounded so similar, giving it the clinician's feel of a genuine phenomenon. The earliest medical reference, a case of sound-induced color, dates to 1710, but the style and details of Galton's report make his the first recognizably modern one. For example, synesthetes express astonishment at discovering that they are unusual, assuming "everybody did it." Most claim to have had it as far back as they can remember and, far from trying to appear special or call attention to themselves, genuine synesthetes prefer to hide their trait because of the ridicule they suffer upon disclosure.

The experience of synesthesia is clearly ineffable, as witnessed by the collective struggle to convey exactly what is sensed. Even computer animations are said to be only about 60 percent representative of "what it is really like." As Galton noted in his 1883 *Inquiries into Human Faculty*, those with visual synesthesia are "invariably most minute in their description of the precise tint and hue of the color. They are never satisfied, for instance, with saying 'blue,' but take a great deal of trouble to express or match the particular blue they mean."

Indeed, one can take the details of a given synesthete today and find matching examples in the classical literature, linking the efforts of scientists a century ago with contemporary ones. Ironically, it is precisely synesthetes' subjective claims that now form the basis of today's experiments that address predictions regarding the trait's perceptual reality (Smilek & Dixon, 2002).

CHARACTERISTICS UPHELD

In chapter 10, I distinguished acquired from naturally occurring synesthesia and stressed that nothing needed medical treatment. In order to differentiate the perceptual trait from metaphor, deliberate contrivances like *son et lumière* or odorama, or conceptual ideas of sensory fusion, I distinguished five characteristics: Synesthesia is (1) involuntary and automatic, (2) spatially extended, (3) consistent and generic, (4) memorable, and (5) affect-laden. How have recent experiments supported or refuted these characteristics?

Involuntary and Automatic

Synesthetes claim to hear a certain sound or look at a letter, for example, and then to see a color. "It just happens," they say. How can we demonstrate that they have no control over their experiences? Phenomena called "perceptual grouping" and "pop-out" confirm that the response is automatic. For example, imagine that a group of 2s arranged to form a triangle is embedded in an array of 5s drawn such that the figures resemble mirror images. When told to look for a hidden shape, most of us would take time to hunt down the target triangle buried within the distracting 5s. But a synesthete who sees every numeral as differently colored would immediately see the target pop out of an alternatively colored background. If the perception is involuntary, synesthetes should perform much faster than nonsynesthetes—and they do.

Because synesthetic associations are idiosyncratic, such tests must be tailored to the individual. That is, two individuals with the same kind of synesthesia will rarely agree as to the particulars of what they perceive. The numeral 2 may be green or red or turquoise for different people. Deliberately inducing mismatches—say by printing graphemes in ink colors that are either congruent or incongruent with a given synesthete's perceptions—and then measuring reaction times to them has become a popular approach in current research. This is called the "modified Stroop paradigm."

In another setup, surrounding a target grapheme in the visual periphery with other letters renders it "invisible," meaning that it is not consciously perceived. Remarkably, it still evokes the synesthetic color. "It must be 'A' because I see red," a subject will say. This implies that synesthesia is evoked at an early sensory level—a preconscious one in fact (Ramachandran & Hubbard, 2001).

As these examples show, many of the probes designed to test if synesthesia is automatic also turn out to prove that synesthesia is perceptual. What are called random dot stereograms do even more, helping us identify the lowest brain level at which synesthesia can occur. When the left eye looks at one pattern of black dots and the right eye at another, the two images fuse in the brain, causing a three-dimensional object to pop out from the viewing plane. Synesthetes see the object, as everyone else does, but they see it in color. This result says two things: that synesthetic color arises after binocular fusion (setting the lower brain limit above the first synaptic level of visual neurons in the cerebral cortex, called V1), and that color appears to be bound to a form as the form is recognized (Palmeri et al., 2002).

Spatially Extended

I now refer to synesthetic percepts as "spatially extended" rather than "projected." Some synesthetes describe Technicolor reading "on the page," even as they simultaneously see the black ink of the printing. Others with colored hearing speak, for example, of watching "a screen about 6 inches from my nose." Michael Watson often reached out in front of him to feel shapes at arm's length. Even those who say the synesthesia is in their "mind's eye" remark that it differs from ordinary vision and imagination by its quality of Euclidean locus, meaning having a sense of physical place. That is, synesthetes speak of "going to" or "looking at" a certain place to examine a sensation. The terms "projector" and "associator," coined by Canadian researcher Mike Dixon, respectively distinguish synesthetic percepts that are outside the body from those that are not.

This quality of spatial extension is particularly dramatic in the perception of what are called *number forms*. (The term is something of a misnomer given that number forms concern not just integers but any concept involving serial order.) The perceptual qualities of spatial location, shape and, often, color become synesthetically joined to semantically ordered concepts such as integers, months, the alphabet, shoe sizes, temperature, and what have you. For example, each day of the week or month of the year may be associated with a different colored shape, which is perceived in a location specific to the individual. Number forms are usually colored and create circles, zigzags, loops, and various tortured configurations. I illustrate many examples in the textbook *Synesthesia: A Union of the Senses*.

Note that we usually speak of synesthesia as "joined senses"—a sound being coupled to a visual photism, for example—but spatial configuration, letters, words, days of the week, and the like are not senses at all; they are categories of knowledge. Because they reckon among the most frequent manifestations of synesthesia, we need to enlarge the definition beyond pure sensory-sensory pairings to include the binding of sensory fragments (qualia) to categories of mental concepts. I will return to this later.

Consistent and Generic

Once established in childhood, synesthetic associations remain stable throughout life, as demonstrated long ago by test-retest situations spanning many years. In a recent demonstration, synesthetes were asked to indicate their color responses to a list of words. When tested without warning a year later, they report almost identical responses, whereas controls without synesthesia, even if forewarned of retesting a month before, perform near chance level.

Synesthetes often remark that some colors they see are "weird," meaning ones that they would never deliberately choose. They may see colors that they do not like, or wish that they saw their favorite ones more often. This should not surprise us, given that their visual systems are being stimulated by nonoptical means over which they have no control. In one interesting example, a color-blind synesthete with S-cone deficiency—which makes it hard for him to discriminate blues and purples—speaks of seeing numbers in "Martian colors," meaning colors he is unable to see in the real world. Curiously, synesthesia happens to be more common in blind individuals than the general population. Moreover, synesthetes have normal color vision (as determined by Isihara plates) and color naming.

Saying that synesthesia is generic, as well as consistent, means that what is experienced is not complex and pictorial, but elementary—blobs, lattices, cold, rough, sour, zigzags, simple geometric shapes, and so forth. They are more *categorical* than particular. This may have to do with constancy operations performed by the brain, which I will return to.

Memorable

When asked what good the trait does, synesthetes immediately answer, "It helps you remember." They do have measurably high memories, sometimes photographic ones, or "eidetic" in psychological parlance. In *Bright Colors, Falsely Seen*, Kevin Dann (1998) considers the relationship between synesthesia and eideticism at length, including the eideticism of synesthete Vladimir Nabokov. There also appears to be an association between synesthesia and perfect pitch. I comment on the two in Cytowic (2002).

Synesthetes say that the extra bits of information help them remember telephone numbers, names, and such. As WW, a synesthetic neuropathologist, puts it, "I use it . . . to help me remember correct sequences of numbers, words, phrases, letters, to help me remember names and locations of anatomical structures (especially neuroanatomical structures—you should see the beautiful array of colors in the brain!) and neuropathological classifications. I could go on and on."

Affect-Laden

Synesthesia carries a sense of certitude, sometimes a "Eureka!" feeling. Most find it highly pleasurable and believe that to lose it would be odious. Trivial tasks are laden with emotional affect, so that mental calculations are "very pleasurable," or recalling a phone number is "delightful." Mismatched perceptions can be "like fingernails on a blackboard."

In a minority of cases, what is perceived is so wretched—for example, vile-tasting words, nausea on playing a musical instrument—that the condition interferes negatively with daily life. Despite this, synesthetes say that they would never part with their perceptions. It is hard to understate the intensity and pervasiveness of affect in synesthesia, and I maintain that no explanation of the phenomenon can be complete without accounting for this emotional charge. Over time, however, it has become apparent that some synesthetes (albeit only a handful thus far) do not feel much affect. This observed variability is theoretically explained by transcription factors, below.

Strong affect aside, the polarity I suggested of a *relative* neocortical deactivation with a corresponding limbic enhancement during synesthetic experience appears overly simplistic in view of new data. I focused on the hippocampus (among other possibilities such as the insula or claustrum) because all sensation converges there together with memory, salience, and affect, thus accounting for all features of synesthesia. It was neat and tidy, but as I have pointed out, nature often makes a mess of neat and tidy mental concepts. I elaborate a new model below that still features a role for affect and salience.

While on the subject of affect, I must confess being surprised at the speed with which the artificial intelligence community has embraced emotion in their work beyond the pioneering work of Paul Werbos. Having bludgeoned strong AI (admittedly a straw man I used for rhetorical purposes), I was delighted to receive Professor Rosalind Picard's invitation to lecture at that high temple of computation, MIT's Media Lab. After playing my assigned role of iconoclast, I enjoyed many hours of demonstrations by her postdoctoral students. These technical exhibitions ranged from medical holography, to musical composition, to computer-generated characters now so familiar from cinema (the twist being that MIT's characters interact with you). Some of their prototype computers are even worn in the form of hats, glasses, gloves, and jackets. Among all this fun, the project that most intrigued me was affective computing.

The term "affective computing" means computation that relates to, arises from, or deliberately influences emotion. Beyond incorporating neural networks that mimic some emotional circuits of the human brain, as Werbos did, affective computing feeds back certain physiological parameters to make the user more aware of his or her physiological and emotional state. This is possible when a computer can recognize and express emotions, respond intelligently to human emotion, and (possibly) "have" and regulate emotions of its own. Affective computers that one wears may yet be the ultimate mood ring.

Affective computing's approach is straightforwardly experiential, meaning physically hands-on rather than mentally logical and unembodied in the sense I use throughout the essays. Professor Picard's approach happily agrees with my own emphasis of experience over explanation. One reason Dr. Picard's project is compelling is that she seeks not to replace human brains with computers but to use them to make humans more attuned to what their brains are doing. Given our frequent physical interaction with computers in everyday life, they are well situated to learn about us intimately, a possibility that has both good and malign prospects.

The information age has brought us perhaps too much information. Affective computers should be able to prioritize, extract what is salient, and manage information overload all the while getting to know—and serve—their users better. Because affective computing projects emotional feedback while engaging multiple senses simultaneously, understanding can occur in more than one way. In my view, this is a mind expansion that no drug can match. During my own hands-on immersion, I perceived a concision of metaphors, a cutting and joining of divergent metaphors into a novel, compound experience. Being physical, the knowledge gleaned from that experience is difficult to put in words (another noëtic understanding). My insight was of a unique kind, a singular qualia. If professor Picard's work succeeds, such noëtic moments may become commonplace. Interested readers can read more about affective computing in Picard (1997) and Evans (2003).

PICTURES, PLEASE

Synesthesia's reality is demonstrated by its automaticity, consistency, and durability; by its induction of perceptual grouping and pop out; by the evocation of colors by "invisible" graphemes at an unconscious level; by the fact that having one type of synesthesia makes one more likely to have a second or third type; by the ability of color-blind and blind persons to see colors that they are unable to sense in real life; and by its strong heritability as an X-linked dominant trait. As a group, synesthetes also appear to have mild neuropsychological deficits in arithmetic, right-left confusion, and navigation.

Despite these kinds of proofs, some skeptics can be satisfied only by machine verifications that produce pictures of the brain. What is remarkable to a clinician like me is how profoundly the emphasis of those pictures has switched from structure to function. When, more than twenty years ago, my colleagues asked about Michael Watson's CAT scan, they expected that a gross brain abnormality must underlie synesthesia if it were real. In other words, where was "the hole in his head"? But given that synesthetes like Michael were normal, manifesting no evident neurological impairment, a structural lesion or bit of missing brain were unlikely. As expected, his CAT and MRI scans, which assess structure, were normal. What was wanted was a test of function.

The cerebral blood flow studies showed that Michael's brain behaved much differently from nonsynesthetic ones, being strongly perturbed by ordinary stimuli such as smell. They also supported predictions from clinical evidence that synesthesia was a phenomenon of the brain's left hemisphere. This left-brain locus disappoints some people, who want it to be a right-brain function because they consider synesthesia artistic and creative. (The assumptions about laterality contained in such a statement are so far off base as to be not even wrong, so I am not going to address it.)

In 1995, Eraldo Paulesu and colleagues performed PET scans on six women who saw colors in response to spoken words. PET offers superior spatial resolution and other advantages to assessing function compared with my technique fifteen years earlier. In this study, spoken words activated auditory and language areas in both synesthetes and controls, but only in the synesthetes did they also activate some visual areas.

Scientists have labeled only a few of the numerous cortical areas involved in vision by a numbering scheme that owes more to animal work and the accident of chronology in elucidating a given area's function than it does to any conceptual logic of how humans see. V1, formerly called the primary visual cortex, is the *first level* at which retinal projections synapse in the cortex. V1 acts like a post office, sorting and forwarding different kinds of signals to different destinations where different types of transformations are carried out, and so it is expected to participate in all visual tasks. At the *second synaptic level*, V5 pertains to motion and direction, V4 to color, and V2 and V3 to form perception. At the *fourth synaptic level*, none of the areas pertaining to facial recognition, facial expression recognition, and spatial-location encoding have yet received a "V" label. Whereas Paulesu's study did not show the hoped-for activation of the unique human color area, V4 (possibly due to a limitation of the PET technique), it did provide another result that was startling: a failure to activate V1 or V2 in synesthetes. These two early visual areas do activate when control subjects view colors.

This result is inconsistent with a major premise of what is called "blindsight." Some brain-damaged patients retain capacities of which they are not conscious. Oxymoronic terms such as "blindsight" or "numbsense" convey how blind or insensate individuals can nonetheless discriminate visual or tactile test targets, despite their insistence that they cannot "see" or "feel" anything. Because stricken individuals are oblivious to their unconscious know-how that allows correct discrimination, researchers have postulated that the primary sensory cortex (such as S1, V1, A1), which is damaged in these individuals, is indispensable for any conscious awareness. In the words of Lawrence Weiskrantz (1997), the acknowledged authority in the field, "striate cortex [V1] is essential . . . for any 'seen' [consciously experienced] perception whatsoever."

Not any longer. Synesthetes in the PET study proved that the brain *can* generate conscious visual experiences without contribution from the primary visual cortex (V1). Blindsight's implications for consciousness studies therefore need to be rethought. In the meantime, synesthesia supports the claim by vision researcher Semir Zeki (1993) that activity in any given module sustaining a given visual function (e.g., V4 for color, V5 for motion, V3 for form) is sufficient, as well as necessary, for one to be conscious of that color, motion, or form. That is, activation of V4 alone is sufficient to "see" color, without the necessity of recruiting other visual modules, either upstream or downstream. (See the "book reviews" page at http://Cytowic.net for my review of Weiskrantz.)

BE CAREFUL WHAT YOU WISH FOR

In 2002, a functional MRI (fMRI) study by Julia Nunn and her colleagues at last confirmed what was long expected: V4 activation (without V1 or V2 activity) in synesthetes who see color in response to spoken words. Whereas both synesthetes and controls activated auditory and language areas as expected, the synesthetes also activated the color area (V4), but only on the left—in agreement with earlier results. Such lateralization is tantalizing, given that their color experiences were not confined to the right visual field. The fMRI technique, which is the most refined one we have to date, also disclosed activation in transmodal areas concerned with memory and affect, consistent with both the subjective statements and clinical observations of synesthetes (table 4).

An unexpected result of this study was that when actually viewing colored surfaces, synesthetes do not activate their left V4, the area for color. Right V4 did function similarly for both synesthetes and controls. Ordinarily, viewing colors activates both right and left V4, as well as the early visual areas V1 and V2. The implication, therefore, is that the participation of left V4 in synesthetic color experience renders it unavailable for ordinary color perception—in other words, synesthesia appears to have hijacked an existing brain function. This surprise is consistent with the observation that nonsynesthetes merely *imagining* colors (compared to performing a visual control task not involving color) do not activate V4. Thus, the brain basis of synesthetic color experience is

Table 4. Summary of activations and deactivations in phonemic lexical synesthesia (spoken words → color), combined from PET (Paulesu et al., 1995) and fMRI (Nunn et al., 2002) studies
No Change
V1, V2
Decrease
Left: insula, lingual gyrus
Activation
Left: V4/V8
Right: inferior prefrontal, insula, superior temporal
Bilateral: posterior inferior temporal (l > r), parieto-occipital junctions

consistent with real color perception rather than color imagery. This refutes earlier criticisms that synesthetes are just "making it up" or have "overactive imaginations."

Lastly, this study has largely overthrown the only plausible alternative explanation of synesthesia that remains, namely that it results from childhood learning through association. This claim said that playing with refrigerator magnets or alphabet blocks, for example, makes some children form enduring associations such as "'A' is red." Rigorous efforts to train controls to imagine colors in response to words demonstrates that this is not so. Despite training until controls achieved 100 percent accuracy, they showed no activity whatsoever in V4 on either side. In a follow-up study that asked whether synesthetes possessed extraordinary associative skills, subjects who had claimed no spontaneous color response to music were trained to associate colors with a melody, as were controls; neither group had activity in the V4 region that had activated when synesthetes heard spoken words. Thus, not only was learning ruled out as an explanation, but also the patterns of brain activity could easily distinguish between subjective states that synesthetes claimed to experience (word-color) or denied having (music-color). Taken together, these results support the existence of a direct neural projection from auditory speech areas to the visual color area known as V4.

Those of us who study synesthesia now mostly concur that inheriting an X-linked dominant genetic mutation results in a failure in synesthetes' brains to prune juvenile projections between brain structures that normally exist temporarily during the development of all brains.

This is what we call the "neonatal hypothesis" for synesthesia: Everyone is born synesthetic, only to lose the capacity as the brain matures. In general, the argument goes like this: (1) All neonates appear to be synesthetic. (2) Many of their neocortical areas function poorly if at all, yet some transmodal entities appear operative. (3) Physiologic necrosis is a normal process whereby transient connections among developing modules form in great excess only to be pruned later. (4) Inheriting a genetic mutation results in failure to prune juvenile connections, and their persistence in mature brains leads to synesthesia. (5) The observation that synesthesia is more common in children suggests that in most individuals neonatal connections are pruned sufficiently so that any "anomalous" hyperbinding among modalities never reaches consciousness.

An exception to point No. 5 above occurs when we are able to quiet the chatter of the cognitive mind. Then we can become aware of our ever-present synesthesia. I suggested in this book that meditation might be such a state, and Roger Walsh at the University of California (2002) has supporting evidence for this conjecture. He finds that synesthesia is one hundred times more common during meditative states compared to baseline prevalence. So, if you wish to experience your inherent synesthesia, do not take LSD but learn to meditate.

With increasing years of experience, the percentage of meditators experiencing synesthesia increases (35 percent vs. 63 percent). Even within the most inexperienced beginners groups, those experiencing synesthesia had twice as much average practice time (17 years) than those who did not experience synesthesia (8 years). Among a third group of adept teachers (who had from 24 to 31 years of practice experience), over half had polymodal experiences and also perceived *categories* synesthetically—thoughts, emotions, and images felt as a sensation, for example. Further relevant observations for all three groups are that synesthesia was most apparent *during* meditation, and that some noted the onset of synesthetic experience only after they had taken up the practice of meditation. Walsh concludes that, "Awareness-enhancing techniques such as meditation may unmask an ever-present synesthesia to consciousness."

Because it is not possible to directly map hardwiring in living humans, we are at present debating precisely where these projections among modalities might lie, and dreaming up ways to confirm or disprove our conjectures. This may take decades, however. In contrast to impressive advances in revealing the functional landscape of human brains, little advance has occurred in understanding its connections. Two levels of neural connectivity contribute to the functional organization of the brain. In the first, genetically determined axonal connections specify the type of information a given area will transform; in the second, experience induces modifications in the synaptic strengths of these connections such that each individual accumulates a unique knowledge base over time.

At present, the objectivists have finally gotten a machine proof of synesthesia upon which they insisted, but at a high cost to their status quo.

CONVENTION UNDER THREAT

The existence of any physical projection as a basis for synesthesia threatens one of contemporary neuroscience's widely held concepts, modularity. As initially proposed by philosopher Jerry Fodor (1983), the mind is constructed of independent subsystems that receive inputs only from a specific category of stimulus and that operate uninfluenced by activity in other modules or systems. The concept of modularity originally referred to cognitive domains, but over time has encroached into the physical organization of the brain, such that relatively self-contained entities such as V1, V4, the grapheme area, and so forth are also referred to as "modules." The mental and physical concepts are not wholly comparable, however, but this is not central to my point. Fodor (2001) has recently abandoned the idea of self-contained modules for cognitive skills (such as language or memory), but continues to promote it for sensory and motor channels. Synesthesia obviously raises the question of whether the concept of modularity *per se* remains entirely valid.

Another endangered favorite of philosophers and cognitive scientists is functionalism. This concept relates to what is called the "hard problem of consciousness," namely, the subjective aspect of perception. Functionalism describes the relations among sensory inputs and their neural transformations, the resulting behavior, and our conscious experience. The concept has engendered many varieties of philosophical argument. One popular formulation states that each subjective experience ("quale," plural "qualia") is identical to the function with which it is associated. That is, functionalism replaces any supposition that red "feels like" a certain state with observable behavior, such as a person saying "red" or pointing to it. Functionalism says that qualia *are* the functions (input–transformations–behavioral output) by which they are supported and nothing more.

If so, then two conditions incompatible with functionalism would be (a) two qualia produced by a single function, or (b) two functions producing the same quale. In synesthetes, the quale of "red," for example, can arise either by optical or by nonoptical routes, producing an example of condition (b) given that two different neural processes on opposite sides of the brain, one optical and the other synesthetic, are both subjectively experienced as the color red.

Another argument put forth for functionalism is that functions giving rise to qualia must benefit the organism, because evolution selects for traits favoring survival. If this is correct, one should not encounter qualia that interfere with the functions of which they are part. However, I have already mentioned the situations where a perceptual mismatch slows performance, to say nothing of the unpleasant and sometimes disruptive affect accompanying some synesthesiae. Other examples of sensory interference appear in the textbook. Nor is there any positive evidence that the quale of color helps aural or visual word perception; in fact, it impedes. These observations are incompatible with the evolutionary claim of functionalism.

In 1997, Jeffrey Gray was the first to notice the danger that synesthesia posed to the hard question of consciousness and has studied this problem in depth (Gray et al., 2002). The arguments outlined here are essentially his. Because functionalism purports to be a general account of consciousness, a single negative instance that it cannot explain is sufficient to render it invalid, just as the axiom, "All swans are white" can be invalidated by observing a single black swan. If functionalism does not work in synesthesia, it does not work anywhere and thus cannot be a general account of consciousness.

The ready objections that synesthetes are not "really" seeing red— that they are merely being artistic or metaphorical, are on drugs, or say what they do only because of a vivid memory for some past association such as refrigerator door magnets—have already been addressed. Because it is unlikely that philosophers will now succeed in eliminating synesthesia, they must either eliminate functionalism or refine it. So far as I know, philosophers never tire of arguing, so I feel confident they will choose the latter course.

Lastly, synesthesia deals a blow to the staunchest objectivists by showing clearly how perception is not passive, how it is not an impression in the brain transferred by objective physics in the world "out there" (philosophers call this direct realism). When a synesthete responds to the word "butter" by saying "blue circles moving off to the right," she demonstrates a lack of correspondence, let alone an identity, between the physical world "out there" that produces the percept and the percept itself. Many other approaches have reached this view that perception is active and constructive; synesthesia happily provides a very clean example.

So much for the wrecking ball. What issues might synesthesia illuminate? Two big ones are the so-called binding problem and metaphor.

THE BINDING PROBLEM

Diverse perceptual attributes (such as color or shape) are processed in different areas of my brain, yet I perceive an apple as a unitary entity, not something red + round + edible + over here, and so on. What is more, attributes are processed not only in different locations, but also at different times in my brain. For example, color is perceived before orientation, which is perceived before motion, with approximate lag times of 30 ms and 40 ms, respectively. Some associations are so firmly bound that we cannot help but perceive the voice as coming from a dummy whose mouth moves, which is why even bad ventriloquists can pull off this illusion. Dance is another example of tight binding between sound and movement. Why is it so hard to sit still when we hear a good beat?

How sundry, asynchronous attributes become bound into a seamless perception—red apple—endlessly baffles neuroscience. It is clear that attributes do dissociate: Consciousness can be impaired solely in highly restricted domains as seen, for example, in such states as achromatopsia, akinetopsia, or the agnosias. Inasmuch as synesthesia binds perceptual qualia in anomalous combinations, might it not say something useful about binding in general?

There is a further twist. I mentioned earlier that synesthesia's most common manifestation is a coupling of sensory qualia to categories of knowledge: for example, color, flavor, texture, locus, or configuration may be bound to letters and integers, members of a serially ordered set (such as days of the week), words, or even symbols such as Braille. Consider how many neurological syndromes (the agnosias) as well as imaging studies demonstrate that we think in categories. In prosopagnosia, for example, stricken individuals can no longer recognize faces. They recognize a face as a face, but cannot say *whose* face it is. Their larger failure is in comprehending examples within a category. Thus, despite all their previous knowledge and skill, a stricken bird watcher says that all the birds look alike, a farmer can no longer distinguish his cows, and a gardener cannot tell one plant from another.

Might synesthesia relate to the brain's search for constancy and the assignment of essential features that constitute a category? An enduring puzzle of neuroscience is how, out of a constantly changing and infinite energy flux, the brain—whose resources are finite—assigns objects their constant features.

Color and form, so prominent in synesthesia, are properties constructed by the brain through what are called constancy operations. For example, most people accept the explanation that something looks red because it reflects red wavelengths more than others, but color is actually a property of brains and not the physical world (for dramatic proof, see Edwin Land's "color Mondrian" experiments in chapter 9 of *Synesthesia: A Union of the Senses*). For surface colors to be perceived as constant despite ever-changing illumination, it is precisely the wavelength composition of reflected light that the brain must ignore. A banana stays yellow whether we view it indoors or outside, in bright sunlight or shade, even though the wavelength composition of these light sources differ enormously. Likewise, all constructed properties require that the brain discount certain things. With color, it is wavelength composition of reflected light that the brain must ignore; with form, it is the viewing angle; and with size, it is viewing distance.

As with color, forms do not exist without a brain. Individuals born blind and to whom vision is later restored find it impossible to learn to *see* even a few forms. There is a critical period after birth during which the genetically determined visual apparatus must receive visual stimulation. Picasso and Rousseau famously said they wished to paint with "the naiveté of a child who has not yet learned to see analytically." But that is Romantic pap, given that the youngest child's brain has already undergone extensive visual training. The neo-Platonist Plotius remarked that, "The form is in the sculptor long before it ever enters the stone." In neurological terms, the Platonic Ideal equals the brain's memory of all the views of all the objects it has ever seen, from which it establishes canonical features.

My discussion of the form constants in chapter 15 raises the question of what their possible neural basis might be. Regrettably, little modern research addresses this issue. A large chunk of visual brain, especially V2 and V3, contains orientation-selective cells that are themselves highly ordered. For example, an electrode penetration perpendicular to the brain surface finds that cells responding to the same orientation are stacked in a column, whereas a sample taken at 45° encounters cells whose preferred orientation gradually changes in an orderly progression. We do not yet know how complex forms are constructed from simpler components that are physically ordered themselves. We must be patient and await further data.

Synesthesia has led me over time to favor a model of brain organization called the distributed system. I can only summarize it here; those wishing greater detail can consult chapter 6 in *Synesthesia: A Union of the Senses* or Mesulam (2002). The prime features of this model are a distribution of function (hence the name) across structures—as in neural networks—and simultaneity of activity on several levels, compared to the more familiar sequential cascade in which a module is assumed to complete its transformation of neural inputs before passing the result on to the next module in the sequence. This older idea may be likened to stations in a factory connected by a conveyor belt by means of which one thing after another is added, whereas the distributed system is like different authors simultaneously writing separate chapters of a book without fully knowing how the other chapters end. The distributed system also departs from the older idea of a strict one-to-one mapping of function to anatomy, depending instead on topological relations and convergent-divergent relations among brain entities. These two features result in the multiple mapping of a given function, as seen in the numerous modules pertaining to vision, some of which we understand better than others (such as V4 for color, or V5 for motion and direction). Relevant to synesthesia, what are called transmodal modules (meaning "not pertaining to any single sense") do three things: They construct multisensory representations of the world, they lend memory and affect to experience, and they critically participate in establishing categories via groups of coarsely tuned neurons.

This model organizes brain tissue into five major networks and many lesser distributed systems. In any one such system, a given cerebral module participates in more than one cognitive function and connects with several-to-many other nodes. A given function is not so much localized in the sense of classical neurology, but exists as the *dominant process* within its distributed system at any given time. Multiple synaptic levels are active simultaneously, each node influencing the state of adjacent levels (as in the example of our simultaneous authors). Such organization reminds us that localization is a function of probability—and not just in this model but in any scheme of neural organization. (Try drawing the boundaries of Wernicke's area on a standard brain atlas—you cannot. See Bogen & Bogen, 1975.) Scans mislead us by emphasizing peak probabilities, which we misconstrue as fixedly anatomical. The answer to synesthesia will not be a "where" but a "what."

It would thus be wrong for me to leave the impression that V4 is the seat of synesthesia: Any module found active by a scan (or other means) is really just one node in the distributed system underlying its expression. The totality of synesthetic experience involves more than the conscious perception of a single quale, as I hope I have successfully conveyed. My comments above regarding the participation of transmodal modules in synesthesia are not incompatible with the idea, mentioned earlier, that an inherited genetic mutation causes extraordinary, one-way projections between cerebral modules that underlie very specific functions. However, a connection between, say, the grapheme area that allows one to understand written numbers and the V4 color area does not fully explain synesthetically colored numbers because it leaves out the affect of the experience, its memorability, whether the synesthetic color moves, has a given spatial location, and so forth. Such a cross-activating projection says nothing whatsoever about all the other kinds of synesthesia; it fails as a general explanation.

As Ramachandran (2001) suggests, what are called transcription factors can partly solve this shortcoming because transcription factors can cause a gene's effects to be expressed either discretely or diffusely—or anywhere between—in the brain. Such variability goes a long way toward explaining the observed variety of synesthetic experience, and why some people have only one kind whereas others have three or four different kinds of synesthesia. Thus, transcription factors expressed in different places throughout the brain could account, theoretically at least, for subsidiary features of synesthesia such as memorability and affective charge. But it is precisely this necessity of widespread expression that makes me point out why synesthetic experience *per se* cannot be localized to any one physical spot in the brain and why scans mislead us badly in this regard.

METAPHOR AND LANGUAGE

The heterogeneity to which I referred above connotes more than a wide variety of perceptual combinations. There is also heterogeneity in the depth of subjective experience, from purely sensory-sensory, to categorical-sensory, to verbal-sensory. In this last, even a concept—just thinking of the number 5, say, or a person named Marion—is sufficient to trigger synesthesia. Some time ago, both Lawrence Marks (1978) and I proposed a cognitive continuum extending from perception to synesthesia to metaphor to language. With time, others are starting to concur.

Systematic correspondences exist among dimensions of a given sense for synesthetes and nonsynesthetes alike. For example, both say that louder tones are brighter than soft tones, that higher ones are smaller than lower ones, and that low tones are both larger and darker than high ones. The perceptual similarities that yield such orderly relationships among pitch, loudness, brightness, and size, for example, turn out to be rooted in fundamental similarities of physical experience itself. Perceptual similarities, synesthetic equivalences, and metaphoric identities in turn become available to the more abstract knowledge that is embodied in language. As I detailed in "The Experience of Metaphor" essay, the acquisition of metaphor relies not on a capacity for verbal abstraction, as many mistakenly believe, but on our physical interaction with the world. Throughout the essays, I explore turning the subjective-objective dichotomy of experience into a unity by showing how we need both points of view.

Objectivity fails to see how the human system of concepts is metaphoric, involving an imaginative understanding of one thing in terms of another. We elaborate the metaphor "The mind is an entity" into another metaphor, "The mind is a machine," when we say, "He ran out of steam." Metaphors emphasize some aspects of an object but hide others. The machine metaphor paints the mind as having a source of power, an on-off state, and an expected level of efficiency, but it hides the vagaries of thought, its ability to make sense of fragmentary information, and the unexpected suddenness of insight. By switching metaphors, we alter how we comprehend a thing.

Subjectivity fails to see that even the most imaginative flights occur in a context of objective experience gained by living in a physical and cultural world. Increasingly, science is viewing metaphor as an emergent property of mind that is rooted in the body. As semiotics have long known, meaning inheres in affect, which the body feels as physical and the mind apprehends as mental. Because metaphor perceives the similar in the dissimilar, it also points to constancy and categorization, features germane to synesthesia. Perhaps a tendency to map one concept to another unconventionally even underlines synesthetes' creativity.

One implication of a continuum from perception to synesthesia to metaphor to language is that synesthesia resides universally in each of us but, for reasons yet unknown, rises to consciousness in only a few. Heinz Werner suggested as much in the 1930s but, as I have demonstrated here, technology takes time to catch up with ideas. Two bits of recent work support this conjecture. One is the study by Roger Walsh finding that synesthesia is one hundred times more frequent during Zen meditation; the other from Madison, Wisconsin demonstrates the ability of both blind and sighted persons to "see" video impulses fed into an electrode array placed on the tongue (Bach y Rita et al., 1998). We do not see with our eyes, anyway, but with our brains. What this latter demonstration shows is that tactile sensations on the tongue can be unconventionally bound to discern form, movement, direction, spatial location, and other qualia that we conventionally ascribe to vision. The capacity for anomalous binding, which is the essence of synesthesia, is therefore latent in all brains.

Nature reveals herself through exceptions. Those objectivists who tried to dismiss synesthesia throughout its history seemed to have forgotten this maxim. Far from being a mere curiosity irrelevant to real issues, synesthesia turns out to illuminate a wide swath of mental life and forces us to rethink some fundamental issues regarding mind and brain. At present, I can think of no issue more relevant to our quest for self-understanding.

▲ ■ ●

Popular appeal has overwhelmed even the impressive revival of scientific interest in synesthesia. Happily, scores of once-anxious, isolated people, most of whom I have never met, now have a name for their experience and the relief of knowing that they are neither alone nor insane. This knowledge has not only alleviated personal unhappiness but also prevented a fair amount of medical mischief.

For example, a man born profoundly hard of hearing wrote to ask if there were any relation between synesthesia and his worsening migraines that seemed to amplify his synesthesia to unbearable levels. "No one can help me," he pleaded, "and I am fed up with close-minded and hardheaded doctors bent on giving me tranquilizers when I tell them about it." He calls his experience "photonic hearing," meaning that the color, intensity, position, and movement of *visual objects* determine the "tonal quality" of the *sounds* they emit. "I am deaf without my hearing aids," he writes, "but silence I have never known. My eyes are another pair of 'eardrums' to me." Who could ignore such frustration, let alone the puzzle of explaining how his brain—already wired differently in utero because of his deafness—produces a synesthesia essentially opposite that of typical colored hearing? I have written in my neuropsychology textbook (Cytowic, 1996) about the similarity of migraine photisms to synesthetic ones in my exploration of form constants and unusual neurological experiences.

I have discussed synesthesia with eager audiences ranging from high schoolers to NASA engineers, from artists to psychologists, from writers to filmmakers to sci-fi fans. *The Man Who Tasted Shapes* exists in German, Japanese, and Korean translations, and the BBC and other companies have filmed documentaries. Synesthetes have formed the International and American Synesthesia Associations (ISA, ASA) and, most importantly, The Synesthesia List. These disseminate information through meetings and on-line resources, and unite synesthetes with one another and with interested researchers in various locales.

These various on-line resources are located at http://Cytowic.net.

References

Bach y Rita, P., Kaczmarek, K., Tyler, M., Garcia Lara, J. 1998. "Form perception with a 49 point electrotactile array on the tongue," *Journal of Rehabilitation Research and Development* 35: 427–430.

Bogen, J. E., Bogen, G. M. 1975. "Wernicke's region—Where is it?," *Annals of the New York Academy of Science* 280: 834–843.

Cytowic, R. E. 1995. "Synesthesia: Phenomenology and neuropsychology," *Psyche: An Interdisciplinary Journal of Research on Consciousness*. [On-line serial at http://psyche.cs.monash.edu.au/v2/psyche-2-10-Cytowic.html.]

Cytowic, R. E. 1996. *The Neurological Side of Neuropsychology*. Cambridge, MA: MIT Press.

Cytowic, R. E. 2002. *Synesthesia: A Union of the Senses*, 2d ed. Cambridge, MA: MIT Press.

Dann, K. T. 1998. *Bright Colors, Falsely Seen: Synesthesia and the Search for Transcendental Knowledge*. New Haven: Yale University Press.

Damasio, A. R. 1994. *Descartes' Error: Emotion, Reason, and the Human Brain*. New York: Putnam.

Evans, D. 2003. *Rethinking Emotions*. Cambridge, MA: MIT Press.

Fodor, J. A. 1983. *The Modularity of Mind*, Cambridge, MA: MIT Press.

Fodor, J. A. 2001. *The Mind Doesn't Work That Way: The Scope and Limits of Computational Psychology*. Cambridge MA: MIT Press.

Galton, F. 1883. *Inquiries into Human Faculty and Its Development*. London: Dent.

Gray, J. A., et al. 1997. "Possible questions of synesthesia for the hard question of consciousness," pp 173–181 in S. Baron-Cohen & J. E. Harrison (eds.), *Synaesthesia: Classic and Contemporary Readings*. Oxford: Blackwell.

Gray, J. A., et al. 2002. "Implications of synaesthesia for functionalism: Theory and experiments," *Journal of Consciousness Studies* 9(12): 5–31.

Marks, L. E. 1978. *The Unity of the Senses: Interrelations Among the Modalities*. New York: Academic Press.

Mesulam, M–M. 2002 *Principles of Behavioral and Cognitive Neurology*, 2d ed. New York: Oxford University Press.

Moldoveanu, M., Nohria, N. 2002. *Master Passions: Emotions, Narrative, and the Development of Culture*. Cambridge, MA: MIT Press.

Nunn, J. A., Gregory, L. J., Brammer, M., et al. 2002. "Functional magnetic resonance imaging of synesthesia: activation of V4/V8 by spoken words," *Nature Neuroscience* 5(4): 371–375.

Palmeri, T. J., Blake, R., Marois, R., et al. 2002. "The perceptual reality of synesthetic colors," *Proceedings of the National Academy of Sciences (US)* 99(6): 4127–4131.

Paulesu, E., Harrison, J., Baron-Cohen, S., et al. 1995. "The physiology of coloured hearing: A PET activation study of colour-word synaesthesia," *Brain* 118: 661–676.

Picard, R. W. 1997. *Affective Computing*. Cambridge, MA: MIT Press.

Ramachandran, V. S., Hubbard, E. M. 2001. "Synesthesia: A window into perception, thought, and language," *Journal of Consciousness Studies* 8(12): 3–34.

Smilek, D., Dixon, M. J. 2002. "Towards a synergistic understanding of synaesthesia: Combining current experimental findings with synaesthetes' subjective descriptions," *Psyche* 8(01), January 2002. [On-line serial at http://psyche.cs.monash.edu.au/v8/psyche-8-01-smilek.html]

Walsh, R. 2002. "Can synesthesia be cultivated? Indications from a survey of meditators." Submitted.

Weiskrantz, L. 1997. *Consciousness Lost and Found*. Oxford: Oxford University Press.

Zeki, S. 1993. *A Vision in The Brain*. Oxford: Blackwell.

Notes

▲ ■ ●

NOTES FOR CHAPTERS 1-9

1. This typical attitude of American medicine starts early in training. Students want to experience exciting crisis intervention and are less interested in preventative medicine. Today, few insurers pay for preventative medicine. Instead of paying a little for vaccinations, well-baby visits, nutritional counseling, screening exams, and efforts to promote healthy life-styles, we pay a fortune to treat advanced end-stage disease, and the maladies associated with tobacco, alcohol, and pollution.

2. Cytowic, R. E. 1976. "Aphasia in Maurice Ravel," *Bulletin of the Los Angeles Neurological Societies*, 41:109–114.

3. This, in fact, was the famous motto of the Royal Society of London, whose charter was granted in 1662 by Charles II, making it the oldest and most venerated of English scientific societies. Modern people often mistranslate the Latin *Nullius in verba* literally as "there is nothing in words," implying that talk is cheap and theories irrelevant. We go astray by misreading the genitive singular *"nullius"* as the nominative *"nullus."* We also fail, as no educated 17th-century person ever would, to recognize the motto as an abbreviated allusion to a longer statement from Horace's *Epistulæ*:

Nullius addictus iurare in verba magistri,

quo me cumque rapit tempestas, deferor hospes

(I am not bound to swear allegiance to the word of any master,

Where the storm carries me, I put into port and make myself at home).

Thus the motto advocates freedom of thought and action, not the insignificance of words. The sense is that learned people would henceforth replace dogmatic philosophical musings with empirical facts and experiments that anyone could reproduce for themselves and see what was true.

4. See Cytowic, 1989. *Synethesia: A Union of the Senses*, pp. 5–10, for a discussion of why medicine had an overhelming indifference to behavior well into this century.

5. Luria, A. R. 1968. *The Mind of a Mnemonist*. New York: Basic Books.

6. For the history of this discipline, see chapter 1 in my textbook, *The Neurological Side of Neuropsychology*, 1996. Cambridge, MA: MIT Press.

7. True, there are other reasons for freely using technology, not the least of which is defensive medicine done in response to the threat of frivolous malpractice suits. Even restrained technological medicine is difficult to achieve, economically undesirable, and possibly professional suicide. Legal, social, and administrative systems all encourage the use of more technology. But I do not wish to go off on tangents away from my main argument.

8. Cytowic, R. E. 1981. "The Long Ordeal of James Brady," *New York Times Magazine*, September 27.

9. The word often means a high knowledge of spiritual things, esoteric or transcendental knowledge. The philosopher Immanuel Kant referred to it as going beyond *a priori* categories. In its original use in Aristotelian philosophy, it meant transcending or existing beyond the bounds of any single category. By the seventeenth century, it was often made synonymous with the term "metaphysical." In uses derived from its philosophical sense, therefore, the word means beyond the limits of ordinary experience. The word "noëtic" also conveys this sense of inner knowledge.

10. Cytowic, R. E., Wood, F. B. 1982. "Synesthesia I: A review of theories and their brain basis," *Brain and Cognition* 1:23–35. Also Cytowic, R. E., Wood, F. B. 1982. "Synesthesia II: Psychophysical relationships in the synesthesia of geometrically shaped taste and colored hearing," *Brain and Cognition* 1:36–49.

11. Locke, J. 1690. *An Essay Concerning Humane Understanding*. London: Bassett. Reprinted 1984 Oxford: Clarendon Press.

12. Newton, I. 1730. *Optiks* (4th ed., 1952). New York: Dover Publications.

13. Castel, L. B. 1725. "Calvecin par les yeux, avec l'art de peindre les sons, & toutes sortes de pieces de musique," *Mercure de France*, 1725, 2552–2557.

14. Castel, L. B. 1735. "Nouvelles experiences d'optique & d'acoustique," *Memories pour l'Historie des Sciences et des beaux arts*, 1735:1444–1482, 1619–1666, 1807–1839, 2018–2053, 2335–2372, 2642–2768.

15. Darwin, E. 1790. *The Botanic Garden, Part 2, The Lives of the Plants, With Philosophical Notes*. London: J. Johnson, reprinted 1978 New York: Garland Publishers.

16. Goethe, J. W. von. 1810. *Zur Farbenlehre*. Tübingen: J. G. Gotta.

17. Suarez de Mendoza, F. 1890. *L'audition colorée*. Paris: Octave Donin.

18. Argelander, A. 1927. *Das Farbenhören und Der Synästhetische Faktor Der Wahrnehmung*. Jena (Germany): Fischer.

19. Devereaux, G. 1966. "An unusual audio-motor synesthesia in an adolescent," *Psychiatric Quarterly* 40(3):459–471.

20. Plummer, H. C. 1915. "Color music–a new art created with the aid of science. The color organ used in Scriabin's symphony "Prometheus." *Scientific American* (April 10); also Sullivan, J. W. N. 1914 "An organ on which color compositions are played. The new art of color music and its mechanism," *Scientific American* (February 21).

21. At forty-five years of age, Kandinsky had already been a professional artist for fifteen years, had previously trained in law (for which he had been offered a professorship), and was fluent in several languages.

22. Theosophy means wisdom concerning things divine and seeks to derive from the knowledge of religious books, or traditions that are mystically interpreted, a more profound knowledge and control of nature than can be obtained by the methods of Aristotelian or other philosophy. The name theosophy is often specifically applied to the system of Jacob Boehme (1575–1624).

23. Charles Baudelaire's poem *Correspondances* is the seminal document of the Symbolist Movement and makes the claim that synesthesia is the essential poetic gift. I find no biographical evidence, however, to suggest that Baudelaire was himself synesthetic. This raises the tangential issue that most people who have heard of the term synesthesia know it only as a literary trope. While interesting in its own right, the deliberate, volitional, and intellectual act of poetically bringing about sensory fusion is distinct from the involuntary sensory experience that I am discussing. In fact, sound symbolism and other poetic types of synesthesia are its polar opposite. The later discussion of Aristotle's common sensibles will clarify this point.

24. Koestler, A. 1968. *The Ghost in the Machine*, p. 7. New York: Macmillan.

25. In fact, DNA analysis shows that humans are more closely related to chimpanzees than chimps are to gorillas. Yet you would never guess this by looking only at external appearances.

26. See Cytowic. 1989. *Synesthesia: A Union of the Senses*, pp. 286–300. Also see Henderson, S. T. 1977. *Daylight and Its Spectrum*. New York: John Wiley & Sons.

27. It is hard for most people to look beyond the illusion of color constancy. Try wearing a bright red jacket and notice how everything around you takes on a reddish hue. How does the red change as you walk from sun to shade? If you succeed in seeing this reality, hang the jacket up and try

seeing more subtle changes. Now ponder how many people look but never see during their entire lives.

28. Brou, P., Sciascia, T. R., Linden, L., & Letvin, L. 1986 "The colors of things," *Scientific American* 255(3):84–91.

NOTES FOR CHAPTERS 10-15

1. James, W. 1901. *The Varieties of Religious Experience*, p. 343. Reprinted 1990. New York: Vintage Books.

2. I have cheated in this example, because the synkinesis is due more to mechanical reasons than neural ones. The tendons of the last three fingers share a common sheath and this contributes to their joined movements when one of them is flexed. Still, it is a compelling example of synkinesis, especially if you don't have a newborn around the house.

3. Nineteenth-century neuroscientists were very much interested in the lower levels of neural integration rather than high. Sir Charles Sherrington, who wrote *The Integrative Action of the Nervous System* (1906), won the Nobel Prize in 1932 for such work.

4. Readers interested in knowing more about this procedure and how its invention was based, in part, on synesthesia, should see Chapter 6 of Cytowic, 1989, *Synesthesia: A Union of the Senses.*

5. Cytowic, R. E. "Seashore Science," *New England Journal of Medicine* 294:(12), March 18, 1976; "Taste–the Unnecessary Sense?" *NEJM* 308(9):530, 1983; "Alexithymia—Or Stupidity?" *NEJM* 313:53, 1985; and "Post-Traumatic Amenorrhea," *NEJM* 314:715, 1986.

6. Cytowic, R. E., Stump, D. A., Larned, D. C. 1987. "Somatic, Ophthalmic and Cognitive Sequellæ in Nonhospitalized Patients with Concussion," in *Nonfocal Brain Injury: Dementia and Trauma,* ed. H. A. Whitaker. New York: Springer Verlag.

7. Cytowic, R. E. 1990. *Nerve Block for Common Pains.* New York: Springer Verlag.

8. See Chapter 8 (pp. 238–283), "Synesthesia and Art," in *Synesthesia: A Union of the Senses.*

9. O'Keeffe, G. 1987. Catalogue notes, National Gallery of Art, Washington, DC, November 1–February 21.

10. Von Hornbostel, E. M. 1926. "Unity of The Senses," *Psyche* 7:83–89.

11. See Cytowic, 1989. *Synesthesia: A Union of The Senses,* pp. 56–60 and 232–235 for discussions of familial instances and the genetic basis of synesthesia.

12. Nabokov, V. 1966. *Speak, Memory: An Autobiography Revisited.* New York: Dover. (First published in 1951 as *Conclusive Evidence.*) Mention of his mother's synesthesia first appears in "Portrait of My Mother," 1949, *New Yorker,* April 9, pp. 33–37.

13. The example comes from Luria's "S."

14. Marshack, A. 1975. "Exploring the Mind of Ice Age Man." *National Geographic* 147(1):62–89.

15. Siegel, R. K., 1977. "Hallucinations," *Scientific American* 237(4):132–140; Siegel, R. K. West, L. J. 1975. *Hallucinations: Behavior, Experience, and Theory*. New York: John Wiley & Sons; Horowitz, M. J. 1964. "The imagery of visual hallucinations," *Journal of Nervous and Mental Diseases*. 138:513–523; Horowitz, M. J. 1975. "Hallucinations: An Information Processing Approach," in Siegel, R. K. & West, L. J. (eds.), *Hallucinations: Behavior, Experience and Theory*. New York: John Wiley & Sons.

16. Klüver, H. 1966. *Mescal and Mechanisms of Hallucinations*, p. 22. Chicago: University of Chicago Press.

17. Kandinsky, V. 1881. pp. 459–460 in "Zur Lehre von den Hallucinationen," *Arkiv für Psychiatrie und Nervenkrankheiten* 11:453–464.

18. Siegel & Jarvik, 1975. "Drug induced hallucinations in animal and man," pp. 81–161 in Siegel & West, 1975.

19. Adler, N. 1972. *The Underground Stream. New Lifestyles and the Antinomian Personality*. New York: Harper & Row.

NOTES FOR CHAPTERS 16-21

1. For references, see *Synesthesia: A Union of the Senses*, pp. 92–93.

2. Eighty-eight percent of those in my study have memories far better than average.

3. Gengerelli, J. A. 1976. "Eidetic imagery in two subjects after 46 years, *Journal of General Psychology* 95:219–225; Pollen, D. A., Trachtenberg, M. C. 1972. "Alpha rhythm and eye movements in eidetic imagery," *Nature* 237:109–112; Stromeyer, C. F., Psotka, J. 1970. "The detailed texture of eidetic images," *Nature* 225:346–349.

4. Scoville, W. B., Milner, B. 1957. "Loss of recent memory after bilateral hippocampal lesions," *Journal of Neurology, Neurosurgery, and Psychiatry* 20:11–21; also, Corkin, S. 1984. "Lasting consequences of bilateral medial temporal lobectomy: Clinical course and experimental findings in HM," *Seminars in Neurology* 4:249–259.

5. Luria, pp. 24, 25.

6. Brust, J. C. M., Behrens, M. M. 1977. " 'Release hallucinations' " as the major symptom of posterior cerebral artery occlusion, a report of two cases." *Annals of Neurology* 2:432–436.

7. Jacobs, L., Karpick, A., Bozian, D., et. al. 1981. "Auditory-visual synesthesia: Sound induced photisms," *Archives of Neurology* 38:211–216.

8. Miller, T. C., Crosby, T. W. 1979. "Musical hallucinations in a deaf elderly patient," *Annals of Neurology* 5:301–302.

9. Coleman, W. S. 1894. "Hallucinations in the sane associated with local organic disease of the sensory organs, etc., *British Medical Journal* 1:1015–1017. Reducing sensory input makes the target cells in the waiting synapse super-sensitive, either to their own spontaneous firing or to random messages from other pathways. This supersensitivity of sensorily deprived neurons explains why the hallucination is perceived in the faulty blind, deaf, or numb field. But it is not sufficient to explain the synesthesia that occurs in persons with no brain pathology at all.

10. Vike, J., Jabbari, B., Maitland, C. G. 1984. "Auditory-visual synesthesia. Report of a case with intact visual pathways," *Archives of Neurology* 41:680–681. This patient is of extreme interest for five reasons. As measured by sophisticated means (1) his vision was normal, (2) his hearing was normal, (3) manipulating the rate and loudness of the sound stimulus changed the perceived movement and intensity of his photisms (this proves that the synesthesia is locked to the stimulus), (4) he saw the photisms only in the left eye and only when the sound stimulus was presented to the left ear, and (5) synesthesia could no longer be induced after the tumor was removed.

11. Gowers, W. R. 1901. *Epilepsy and Other Chronic Convulsive Diseases*, 2nd ed., London.

12. Hausser-Hauw, C., Bancaud, J. 1987. "Gustatory hallucinations in epileptic seizures: Electrophysiological, clinical and anatomical correlates," *Brain* 110:339–359.

13. Penfield, W., Perot, P. 1963. "The brain's record of auditory and visual experience" (p. 635), *Brain* 86:595–696.

14. Having the levels of your neutotransmitters altered doubtless sounds horrible to many readers. But they do not remain static. Some of the largest changes come after mealtimes and something as ordinary as a cup of coffee can cause a significant change in the ratios of various brain chemicals.

15. To be precise, a mild and diffuse increase in frontal lobe activity is often seen. Years ago, this was considered normal. Now we know it to be an artifact of the experimental setting. The "resting hyperfrontal flow" appears when subjects are excited, alert, and curious about this novel procedure.

16. The actual procedure of the xenon[133] inhalation test and the mathematical analysis of its reams of data gathered is far more complex than I have portrayed. Various equations exist to separate factors from one another and numerous parameters are used to assess the purity of the data and look for possible contamination. Using only the dozen most common parameters on our system's sixteen probes yields 192 separate calculated values, a large data set. Because Michael's patterns were so unusual, Dr. Stump and I scrutinized and cross-checked the results until we were convinced that the data were an accurate reflection of the actual metabolic changes going on in Michael's brain.

17. The decrease in Michael's right hemisphere was a passive reaction to circulatory changes, and unimportant to this discussion.

18. See MacLean, P. 1990. *The Triune Brain in Evolution: Role in Paleocerebral Functions*, New York: Plenum, especially pp. 228–244, for detailed summary of this work.

19. The existence of volume transmission appears to utterly falsify all connectionist theories of the brain, that is, those which depend on components connected by circuits. Connectionists are trying to refute the implications of volume transmission, which is a good example of how a single instance can potentially falsify an entire set of theories. For a nonspecialized review of volume transmission, see Agnati, L. F., Bjelke, B., Fuxe, K. 1992. "Volume Transmission in the Brain," *American Scientist* 80(4):362–373.

20. See for example the work of anatomist Este Armstrong, 1986. "Enlarged limbic structures in the human brain: the anterior thalamus and medial mammalary body," *Brain Research* 394–397; "The limbic system and culture: an allometric analysis of the neocortex and limbic nuclei," *Human Nature* (in press); 1990. "Brains, bodies, and metabolism," *Brain Evolution and Behavior* 36:166–176.

Anatomy shows us how every major division of the nervous system has some physical structure related to emotion. The neocortex has the prefrontal lobes; mesocortex has the cingulate gyrus; archicortex has the hippocampal formation; basal ganglia have the amygdala; diencephalon has the dorsal thalamus and hypothalamus; midbrain has the central gray matter; pons and medulla have nuclei of the integrated autonomic relays; and spinal cord has cell column nuclei. Such a laundry list shows that the nervous system clearly has a central *emotional core*. How can emotion be dismissed when so much brain tissue is devoted to it?

21. This finding is supported by reports that lack of cerebral oxygen can produce hallucinogenic images. See *Synesthesia: A Union of the Senses*, p. 129.

22. Nieuwenhuys, R., Voogd, J., van Huijzen, C. 1988. *The Human Central Nervous System, A Synopsis and Atlas*, 3rd ed. New York: Springer Verlag. Of course, by this time readers will have likely forgotten that it was the limbic system that so enraged me to throw my notes off the balcony as a student.

23. For multiple references to this point, see Cytowic, 1989. *Synesthesia: A Union of the Senses*, p. 174.

24. Kornhüber, H.H., 1974. "Cerebral cortex, cerebellum and basal ganglia: A introduction to their motor function." In F.O. Schmitt & F.G. Worden (eds.), *The neurosciences third study program*, pp. 267–280. Cambridge, MA: MIT press. Kornhüber, H.H., Deecke, L., 1965. "Hirnpotentialänderungen bei Wirlkürbewegungen und passiven Bewegungen des

Menschen: Bereitschaftspotential und reafferente Potentiale." *Pflüger's Archiv für die Gesamte Physiologie*, 284:1–17.

25. See Cytowic, 1989. *Synesthesia: A Union of the Senses*, p. 313, for references.

26. Libet, B., 1978. "Subjective and neuronal time factors in conscious sensory experience, studied in man, and their implications for the mind-brain relationship. The Search for Absolute Values in a Changing World," Vol II, pp. 971–973. Proceedings of the Sixth International Conference on the Unity of the Sciences (San Francisco, November 25–27). International Cultural Foundation Press.

27. Edwin Diller Starbuck was an American psychologist who wrote a classic called *The Psychology of Religion* (1889). This work and Starbuck's manuscript collection are referred to throughout William James' *The Varieties of Religious Experience* (1901).

28. Jiyu-Kennett, P.T.N.H., 1987 *Zen is Eternal Life*. Shasta Abbey Press, PO Box 199, Mt. Shasta, CA 96067. See also Jiyu-Kennett, P.T.N.H. 1977. *How to Grow a Lotus Blossom*. Shasta Abbey Press.

29. These phrases are from the fukanzazengi, the Zazen rules for evening service.

30. The role of the clock in changing civilization is, of course, a favorite theme of the historian Lewis Mumford. See *World Technics and Civilization*, 1963, New York: Harcourt Brace & World.

31. Jeffress, L.A., ed. 1951. *Cerebral Mechanisms in Behavior: The Hixon Symposium*. New York: John Wiley.

32. See, for example, Bogen, J.E. 1975. "Some educational aspects of hemispheric specialization," *UCLA Educator*, 17:24–32. Reprinted in Wihrock, M.C. (ed.), 1977 *The Human Brain*. Englewood Cliffs, NJ: Prentice-Hall.

NOTES FOR PART II

1. Cherniak, C. 1986. *Minimal Rationality*. Cambridge, MA: MIT Press.

2. See, for example, Edleman, G.M. 1989. *The Remembered Present: A Biological Theory of Consciousness*. New York: Basic Books, and Edelman, G.M. 1992. *Bright Air, Brilliant Fire: On the Matter of the Mind, A Nobel Laureate's Revolutionary Vision of How the Mind Originates in the Brain*. New York: Basic Books.

3. Interestingly, the strong anthropic principle is an important part of quantum mechanics, which up to now is the best physics we have to explain the fundamental operations of the universe.

In *The Emperor's New Mind*, for example, Roger Penrose concludes that consciousness is noncomputational and occurs at the level of quantum indeterminacy. This vexes the artificial intelligence camp, who gamble that the mind is some type of mathematical program. Materialists who have discarded dualism argue yet another viewpoint that the unique thing called mind cannot be separated from its physical structure. One then has to call the mind an emergent property of the physical brain or else explain it in terms not reducible to elementary particle physics or chemistry. Penrose resolves these issues by placing mind at the quantum level and promises that when we really understand quantum mechanics we will suddenly understand mind as well. This position is called "promissory" materialism because it claims to be right, but can only promise the necessary proof later. Penrose's analysis is one of many examples of how deeply people from various walks of life are now thinking about consciousness.

4. The word means the "fifth essence," of ancient and medieval philosophy, supposed to be the substance of which the heavenly bodies are composed and to be actually latent in all things. Its extraction by distillation or other methods was one of the great objects of alchemy. It now means the most essential part of a thing.

5. Dyson, F. 1971. "Energy in the Universe," in *Energy and Power*. San Francisco: W.H. Freeman. See also Ommaya, A.K. 1993. "Neurobiology of Emotion and the Evolution of the Mind," *Journal of the American Academy of Psychoanalysis*.

6. Ommaya, A.K. Comments from a symposium organized by him entitled "Consciousness and Computers" at the Smithsonian Institution, August 1991. See also "Neurobiology of Emotion and the Evolution of Mind." *Journal of the American Academy of Psychoanalysts*, 1993.

7. Simon, H.A., Newell, A. 1958. "Heuristic problem solving: The next advance in operations research," *Operations Research* 6 (Jan–Feb):8.

8. Simon, H. 1960. "The Shape of Automation." Reprinted in *Perspectives on the Computer Revolution*, ed. Pylyshyn, Z.W. 1970. Engelwood Cliffs, NJ: Prentice-Hall.

9. Werbos, P.J. 1992. "The Cytoskeleton: Why It May Be Crucial to Human Learning and Neurocontrol," *Nanobiology* 1(1):75–95; "Neural networks and the human mind: New mathematics fits humanistic insight." IEEE Procedings of the 1992 conference on systems, man, and cybernetics.

10. Weizenbaum, J. 1976. *Computer Power and Human Reason: From Judgement to Calculation*, pp. 14–16. New York: W.H. Freeman.

11. Mumford, L. 1963. *Technics and Civilization*, p. 15. New York: Harcourt Brace Jovanovich.

12. Shelley, M.W. 1985. *Frankenstein*. London: Penguin Books. No movie version does Mary Shelley's 1818 novel justice, because they do not allow the monster to speak. In the book, he has many fascinating things to say about his maker and the world he finds himself in. Shelley wrote *Frankenstein* partly in response to circumstances in her new age of capitalist produc-

tion. In the guise of a hero scientist, Shelley also explored the idea of a Promethean maker and shaper of men. Maurice Hindle, in the foreword to this edition (pp. 25–26), says: "Is it not the modern experience of feeling manipulated by forces larger than ourselves (which are nevertheless humanly managed)—Big Science, technology, the 'machinery' of State, international business empires, the mass media, and so on—that links the lay person's predicament with that of Frankenstein's Creature, he who has been put together from dead human parts and then infused with 'a spark of being,' without having any say in the form or purpose of his own genesis?"

13. In *Synesthesia: A Union of the Senses*, pp. 178–183, I discuss the semantic differential, a standard technique for measuring semantic meaning, and how this technique was based on synesthetic research.

14. Standard references on metaphor can elaborate how concepts are grounded, structured, related to each other, and defined. I have particularly relied on *Metaphors in the History of Psychology*, Leary, D. E. (ed.). New York: Cambridge University Press, 1990; *Metaphors We Live By*, Lakoff, G., Johnson, M. Chicago: University of Chicago Press, 1980; and *More than Cool Reason: A Field Guide of Poetic Metaphor*, Lakoff, G., Turner, M. 1989. Chicago: University of Chicago Press.

15. Many of the examples that follow are suggested by Lakoff & Johnson (see above note).

16. Aristotle, ca. 330 B.C. *Rhetoric*, 1410b. Leary (1990) explicates his point to mean that "strange words" are "unintelligible" whereas "current words" are "commonplace." It is between these two extremes of odd speech and cliché that metaphor both pleases and teaches. Poet Wallace Stevens expressed a similar view in saying that "reality is a cliché from which we escape by metaphor." Stephens, W. 1982. "Adagia," p. 179, in *Opus Posthumous*, Morse, S.F. (ed.) New York: Vintage.

17. There is a down side to our innate capacity to discern fundamental patterns, as in form constants and metaphors. Systematic and ubiquitous errors such as imputing order to random phenomena, sharpening the central point of stories while omitting qualifications, and remembering only positive instances while forgetting nonconforming negative ones are examples of human reasoning gone awry called cognitive errors. The very structure of our brain and its concepts frequently distorts what we think is real. See Gilovich, T. 1991. *How We Know What Isn't So: The Fallibility of Human Reason in Everyday Life*. New York: Macmillan.

18. See for example de Sousa, R. 1983. *The Rationality of Emotions*. Cambridge: MIT Press.

19. I am indebted to Dr. Joseph Bogen for bringing Wigan's work to my attention.

20. See for example Weiskrantz, J. 1986. *Blindsight: A Case Study and Implications*. Oxford: Oxford University Press; and Jelicic, M., Bonke, B., et.

al. 1992. "Implicit Memory for Words Presented During Anæsthesia," *European Journal of Cognitive Psychology* 4:71–80.

21. See my comments about Frankenstein, p. 203.

22. The phrase is from philosopher Philip Golabuk, 1989. *Recovering From a Broken Heart.* New York: Harper & Row.

23. Skinner, B.F. 1974. *About Behaviorism*, pp. 234–251. New York: Alfred Knopf.

24. The reference is to a famous paper on this problem by Thomas Nagel: "What is it like to be a bat?" in *Philosophical Review*, vol. 83, 1974. Reprinted in Readings in Philosophy and Psychology, vol. 1, ed. N. Block. Cambridge, MA: Harvard University Press, 1980. See also Nagel's *The View From Nowhere*, 1986. New York: Oxford University Press.

25. Huxley, A. 1946. *Science, Liberty and Peace*, pp. 35–36. New York: Harper and Brothers.

Suggested Reading

▲ ■ ●

Below are some titles on which I have relied, grouped by general topic, and which readers may be interested in exploring further. Detailed references are given in the notes.

SYNESTHESIA

Cytowic, R.E. 1989. *Synesthesia: A Union of the Senses*. New York: Springer Verlag. The first English text on the subject, and also the first to consider it from a neurological as well as psychological point of view.

Luria, A.R. 1968. *The Mind of a Mnemonist*. New York: Basic Books. Luria's account of a memory expert whose art is enhanced by his synesthesiæ.

Messiaen, O. 1956. *Technique de mon Langage Musical*. Paris: Alphonse Leduc. The French composer describes his invention of his famous "modes of limited transposition" as the means by which he renders the colors of his music. Messiaen's synesthesia is mentioned in nearly every biography about him. See also Chapter 8 in Cytowic, 1989. *Synesthesia: A Union of the Senses*.

Nabokov, V. 1966. *Speak, Memory: An Autobiography Revisited*. New York: Dover. The writer's account of his own colored hearing and that of his mother. His mother's synesthesia was first mentioned in a 1949 *New Yorker* profile entitled "Portrait of My Mother" (April 9, pp. 33–37).

REASON AND EMOTION

Cherniak, C. 1986. *Minimal Rationality*. Cambridge, MA: MIT Press. A well-known text challenging the myth of "man the rational animal," as well as the central role that rationality has often been assigned in philosophy, psychology, cognitive science, and even economics.

de Sousa, R. 1987. *The Rationality of Emotion*. Cambridge, MA: MIT Press. An illustration how emotions are partly rational, and an examination of the widespread belief that reason and emotion are natural antagonists.

Tuchman, B. 1984. *The March of Folly: From Troy to Vietnam*. New York: Ballantine Books. Tuchman demonstrates the widespread, but incorrect, view that reason and emotion are antagonistic to each other in asserting that "the prime cause of folly is the rejection of reason." I believe that if the people she cites had understood that their decisions were fundamentally emotional in the first place, they would not have deluded themselves and stepped into folly.

Darwin, C. 1872. *The Expression of the Emotions in Man and Animals*. Reprinted 1965. Chicago: University of Chicago Press. Darwin's classic book is still fresh. He shows that all creatures express some degree of emotion, and suggests that what distinguishes us from animals is not our better developed ability to reason but our better developed ability to express emotion.

OBJECTIVITY AND SUBJECTIVITY

Gardner, J. 1983. *The Art of Fiction*. New York: Bantam. A novelist explicates how the craft of fiction can create a separate reality into which our total awareness is pulled.

Leary, D.E. (ed). 1990. *Metaphors in the History of Psychology*. New York: Cambridge University Press. Explains why we turn to metaphor to describe psychological states such as consciousness, mentation, emotion, motivation, and learning.

Nagel, T. 1986. *The View From Nowhere*. New York: Oxford University Press. How can we transcend our own experience and consider the world from an objective vantage point that is "nowhere in particular"? Nagel warns against excessive objectification.

Weizenbaum, J. 1976. *Computer Power and Human Reason*. New York: W.H. Freeman. A computer-science professor from MIT gives technical and moral reasons why artificial intelligence cannot be built.

INNER KNOWLEDGE AND SPIRITUALITY

Barrow, J. D., Tipler, F. J. (eds). 1986. *The Anthropic Cosmological Principle*. New York: Oxford University Press. The theory that holds that the fundamental structure of the universe is determined by intelligent observers.

Bettelheim, B. 1977. *The Uses of Enchantment*. New York: Knopf. A persuasive revelation of the irreplaceable importance of fairy tales in the education, support, and liberation of emotions in children of all ages.

Campbell, J. Nearly any work by this popular mythologist will do. See, e.g., *An Open Life*, 1988. New York: Larson Publications; *Myths to Live By*, 1973. New York: Bantam; and *The Inner Reaches of Outer Space: Metaphors as Myth and as Religion*, 1986. New York: Harper & Row.

James, W. 1901. *The Varieties of Religious Experience*. Reprinted 1990. New York: Vintage Books.

Jiyu-Kennett, P.T.N.H., 1989 *Serene Reflection Meditation*. Mt. Shasta, CA: Shasta Abbey.

Index

▲ ■ ●

About the Author

▲ ■ ●

Richard Edmund Cytowic, M.D., has authored both neurology textbooks and popular works. He was nominated for the 1982 Pulitzer
Prize for his *New York Times Magazine* cover story about the condition of White House Press Secretary James Brady, who received a
gunshot wound to the brain during the attempted assassination of
President Ronald Reagan.

Dr. Cytowic has appeared often on national and international
radio and television, including *All Things Considered*, *Voice of America*, and *Good Morning America*. His work has been reported in
national publications such as *US News & World Report*, the *Washington Post*, and the *Los Angeles Times*.

Dr. Cytowic received his B.A. in chemistry from Duke University,
and his M.D. from the Bowman Gray School of Medicine of Wake
Forest University. The New Jersey native studied at the University of
London's National Hospital for Nervous Diseases, trained in ophthalmology and neuropsychology, and later served as Chief Resident in
neurology at George Washington University before entering private
practice. He lives in Washington, D.C.

The son of a physician and an artist, he has long been interested
in the harmony between science and art. His medical biographies of
Chekhov and Ravel have won awards. He has several times been a
Resident Fellow at the Hambidge Center and the Virginia Center for
Creative Arts, both Southern artists' colonies.

Dr. Cytowic has been an invited speaker at the World Congress of Neurology, the National Science Foundation, and the American Association for the Advancement of Science. He is listed in *Who's Who in America* and *Who's Who in the World*, serves on the editorial boards of the journals *Brain & Language* and *Brain & Cognition*, and is a Fellow of Britain's Royal Society of Medicine.